The Big Tough Expensive Job

James Laxer
Anne Martin
editors

THE BIG TOUGH EXPENSIVE JOB

Imperial Oil and the Canadian Economy

Press Porcépic

Printed in Canada
1 2 3 4 5 80 79 78 77 76

DISTRIBUTION

CANADA
Musson Book Company
30 Lesmill Road
Don Mills, Ontario
M3B 2T6

U.S.A.
Books Canada Inc.
33 East Tupper Street
Buffalo, New York
14203

U.K. AND EUROPE
Books Canada Ltd.
1 Bedford Road
London N. 2, England
U,K.

Canadian Cataloguing in Publication Data

Main entry under title:

The Big tough expensive job

ISBN 0-88878-122-9 bd. ISBN 0-88878-123-7 pa.

l. Petroleum industry and trade - Canada.
2. Energy policy - Canada. 3. Imperial Oil Limited.
I. Laxer, James, 1941- II. Martin, Anne, 1951-

HD9574.C22B53 338.2'7'2820971 C76-017122-X

Contents

2
FUELISHNESS IN OTTAWA: THE FAILURE OF CANADIAN ENERGY POLICY

3
THE PUBLIC AND THE NATIONAL ENERGY DEBATE

4
PETROLEUM POLICY ALTERNATIVES

Acknowledgements

The work of many people in Toronto and in Ottawa made possible the two conferences of the Public Petroleum Association at which much of the material in this book was presented. We would like to thank all the people who helped plan and organize the conferences, in particular, Abraham Rotstein, Bruce Willson, John Bingham, John Sharkey, Julia Bass, Paul Craven, Sharleen Treleaven, Pat McDermott, Ginnie Stolee, and Erling Stolee.

Introduction

When Canadians thought of Imperial Oil a decade or two ago, images of the Esso dealer on the corner, welcome fuel and service on highways in every part of Canada, and the passion and excitement of the Toronto Maple Leafs and the Montreal Canadiens came to mind. Imperial seemed to be a national institution, guided quietly by a competent Canadian management. Few could remember life without Imperial as the nation's largest oil company. The fact that the company was owned abroad was either unknown or apparently irrelevant. Very much a part of Canada, the Imperial name was synonymous with the oil industry as a whole.

Today the petroleum industry and Imperial bring a new set of images to mind. Concerned about skyrocketing gas and oil prices and impending energy shortages, Canadians are asking tough questions about the record of the petroleum industry in Canada and about the energy policy the government has pursued.

At the centre of the public's concern is the unanswered mystery of Canada's disappearing petroleum reserves. Up until 1973, the petroleum industry told the government that the nation could look forward to oil and gas surpluses for many decades to come. Then the international price of oil quadrupled, and the same petroleum companies changed their predictions. They told the government that without steep increases in the price of oil, the nation would face shortages of oil and gas by the early 1980's. What is remarkable is that the federal government went from the first set of predictions to the second without asking the question that is on the minds of many Canadians: how did it happen?

This book asks that question. It brings together many of the leading critics of the petroleum industry in Canada to probe the seemingly incredible changes in energy policy that have occurred in the past three years. The book takes as its central theme the role of the oil companies in Canada and, in particular, the role of Imperial Oil, the nation's most powerful petroleum company. Subsidiary of Exxon, the world's largest industrial corporation, Imperial has been the pacesetter in advising Canadian governments on petroleum policy. As Canadians know from the company's television advertising on *Hockey Night in Canada*, Imperial Oil is engaged in *the big, tough, expensive job* not only of developing petroleum supply but also of shaping government policy to serve its own ends.

The Big Tough Expensive Job analyzes the power of Imperial Oil and the petroleum industry in general and provides a variety of perspectives on the alternatives facing Canada. This book demonstrates that critics of differing political persuasions are becoming united in the belief that the decisions the multinational oil companies have made for Canada have not been in the national interest and that future decisions must be made by Canadians and must be arrived at publicly and democratically.

It was this same sentiment that inspired the formation of the Public Petroleum Association of Canada last year. The PPAC is a non-partisan organization whose purpose is to advocate a national petroleum policy that serves the interests of Canadians. The sponsors of the PPAC are critics of the petroleum industry as well as Canadians active in politics, the arts, and the labour movement. They believe that an alternative energy policy can only be achieved when the powerful lobby maintained by the oil companies in Ottawa and the provincial capitals has been effectively countered.

Much of this book grew out of two conferences held by the PPAC in Toronto and in Ottawa. In addition to analyses of current petroleum policies and alternatives developed in papers at the two conferences, the book contains personal interviews by the novelist Dave Godfrey with three Canadians who have worked in different positions in the petroleum industry and who have come to the common conclusion that it is difficult to reconcile a commitment to Canada with service to the multinational petroleum industry. The concluding essay, on an alternative petroleum policy, grew out of an address given by Walter Gordon in March, 1976 to a conference on the Politics of Resource Development at the University of Manitoba.

While the critics of the petroleum industry share much in common, they have their own individual points of view and their disagreements. Authors in this book are responsible only for their own contributions. But the facts speak very clearly and the need for thoughtful and major policy innovation is obvious.

James Laxer Anne Martin
Toronto, August 10, 1976

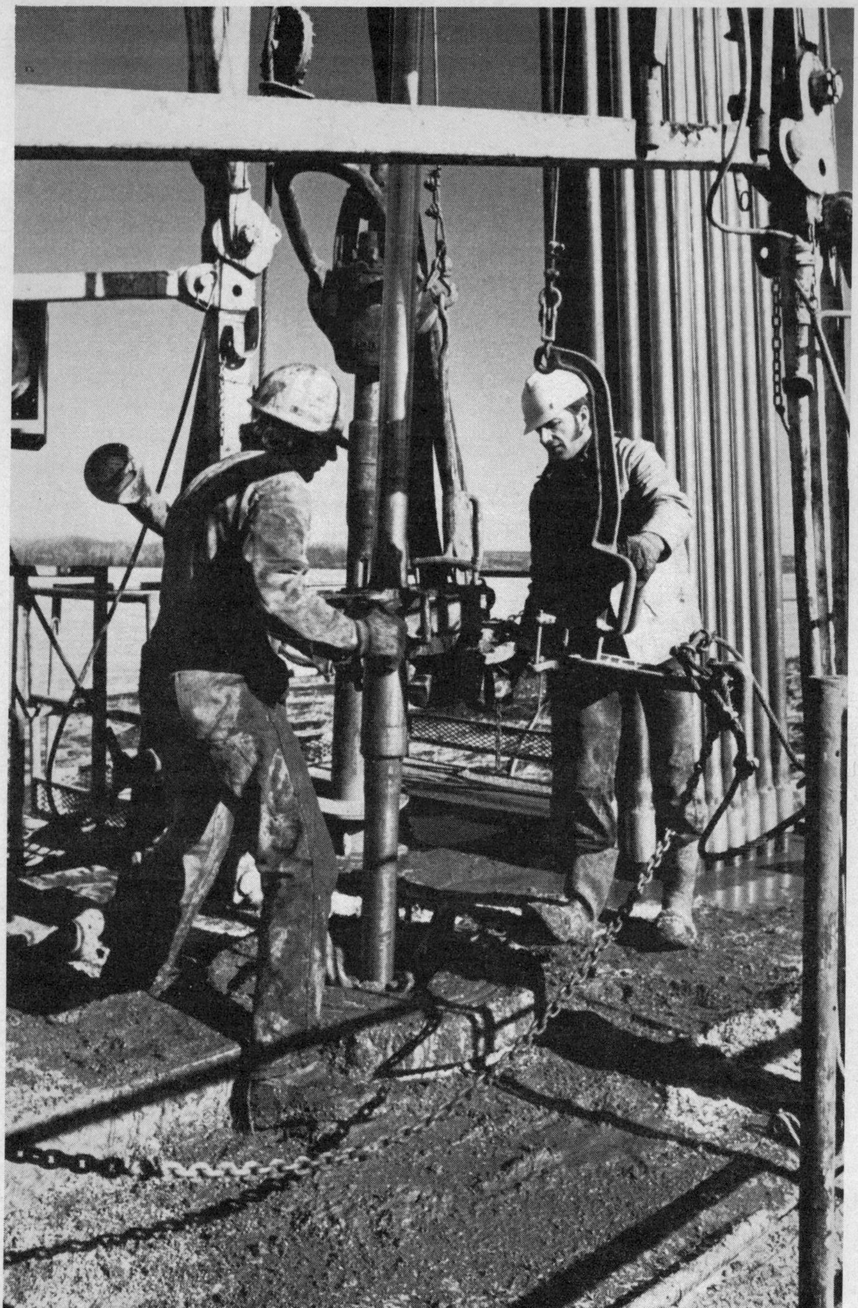

Refinery

1

HAPPY MOTORING: EXXON/IMPERIAL AND CANADIAN ENERGY STRATEGY

EDWARD SHAFFER

Edward Shaffer is a Professor of Economics in the Faculty of Business Administration and Commerce at the University of Alberta. Born in Pittsburgh, Pa., he worked as an economic consultant in New York from 1953-63. From 1962-3 he worked for Walter Levy, the renowned American oil consultant. He is author of *The Oil Import Program of the United States: An Evaluation* (1968), and numerous published articles on Canadian energy policy.

1

Edward Shaffer

The Empire of Exxon

The crucial question facing Canada today is how to develop its energy resources in a way which will maximize the benefits to the Canadian people.

One must have serious doubts as to whether present energy policies accomplish this end. These doubts arise from the leading role played by the privately-owned, multinational companies in the exploitation of Canada's resources. Of these companies, Imperial oil is the most important. Imperial is important not only because of its own size but because of its links with Exxon, the world's largest industrial company.

Like so many other companies operating in Canada, Imperial is a satellite company, owned and controlled by an outside master. Its master, Exxon, owns 70 percent of its stock, dictates its policies and sees that these policies conform to Exxon's worldwide strategy. In this paper I will discuss the worldwide activities of Exxon without going into the details of their impact in Canada.

Exxon is much more than a company. It is an empire with enormous wealth and power. It can be viewed as a microcosm of modern capitalism, in general, and American imperialism, in particular. Let's take a look at this wealth and power.

In 1974, Exxon's worldwide sales exceeded $45 billion (U.S.), the largest of any industrial company in the world. It valued its total assets at over $31 billion and its stockholders' equity at over $15 billion. Its sales in 1974 were larger than the GNPs of most countries in the world. Probably no more than 20 nations had GNPs larger than $45 billion.

Its net income, after taxes, of $3.1 billion was larger than the sales of most of the world's largest companies. Of *Fortune's* 500 largest U.S. industrial companies, only 51, including Exxon, had sales greater than

W.L. Mackenzie King: Former Employee of the Rockefellers

Mackenzie King (centre) with Archie Dennison (left) and John D. Rockefeller (right) while studying industrial relations under the auspices of the Rockefeller Foundation, Colorado, 1915.

Exxon's after-tax profits.

Exxon and its satellite companies operate 70 refineries in 37 countries, including 9 in Canada. They have concessions in the Middle East, Africa, Latin America, North America, and Asia. They own marketing facilities in most countries of the non-communist world. They own and control a large fleet of tankers, pipelines, storage facilities and chemical plants.

How did Exxon acquire such enormous wealth? The economic fiction, too often propogated in our universities, tells us that this wealth represents the company's contribution to social well-being. In a free, market society, so the fiction says, the factors of production are rewarded according to the value of their marginal product. Economic reality tells us otherwise. The accumulation of such wealth came through what we might euphemistically call "strategy" and "planning". In the law of the jungle, which characterizes the operations of the business world, it means utter ruthlessness and the use of highly questionable methods.

It is interesting to note how society tends to put a cloak of respectability on acquired wealth, regardless of the method of acquisition. Today both Exxon and the family which controls it, the Rockefellers, have this image of respectability. But this was not always the case. In the days when John D. Rockefeller was acquiring his fortune, the name Rockefeller brought forth the same connotations as the name Godfather does today. In American politics opposition to Rockefeller was as sacred as the promotion of motherhood. This popular resentment against Rockefeller's tactics culminated in the U.S. Supreme Court decision of 1911, which broke up the old Standard Oil Company of New Jersey into a number of component companies. In delivering the opinion of the Court, Chief Justice White commented specifically on Rockefeller's methods:

> . . . we think no disinterested mind can survey the period in question without being irresistibly driven to the conclusion that the very genius for commercial development and organization which it would seem was manifested from the beginning soon begot an intent and purpose to exclude others . . . which . . . necessarily involved the intent to drive others from the field and to exclude them from the right of trade and thus accomplish the mastery which was the end in view. . .[1]

The defenders of Exxon today, while admitting its questionable past, argue that the present company is socially responsible and has long since abandoned questionable tactics. This may be so. But it is interesting to note that the heirs of John D. have never given up the wealth acquired through these methods. I think we can legitimately question whether Exxon has thrown overboard the spirit of John D. Perhaps it has merely adopted more sophisticated methods to achieve the same end.

The Fortune 500
The Largest Industrial Corporations in the U.S.: 1975 (ranked by sales)

Rank '75	Rank '74	Company	Sales ($000)	Assets ($000)	Rank	Net Income ($000)	Rank
1	1	**Exxon**	44,864,824	32,839,398	1	2,503,013	1
2	2	General Motors	35,724,911	21,664,885	2	1,253,092	3
3	4	**Texaco**	24,507,454	17,262,448	3	830,583	4
4	3	Ford Motor	24,009,100	14,020,200	6	322,700	23
5	5	**Mobil Oil**	20,620,392	15,050,287	5	809,877	5
6	6	**Standard Oil of California**	16,822,077	12,898,150	7	772,509	7
7	9	International Business Machines	14,436,541	15,530,476	4	1,989,877	2
8	7	**Gulf Oil**	14,268,000	12,425,000	8	700,000	8
9	8	General Electric	13,399,100	9,763,500	11	580,800	11
10	11	Chrysler	11,699,305	6,266,728	17	(259,535)	492
11	10	International Tel. & Tel.	11,367,647	10,407,941	9	398,171	15
12	13	**Standard Oil** (Ind.)	9,955,248	9,854,099	10	786,987	6
13	12	U.S. Steel	8,167,269	8,148,174	12	559,614	12
14	14	**Shell Oil**	8,143,445	7,010,753	14	514,827	13
15	18	**Atlantic Richfield**	7,307,854	7,364,787	13	350,395	17

John D., himself, was very cognizant of his poor image. He hired two people to improve it. One was a public relations expert by the name of Ivy Lee and the other was a social-worker turned spiritualist, W.L. Mackenzie King. Judging by the new public attitude toward the Rockefellers—remember that Nelson Rockefeller is still only one heartbeat away from the U.S. presidency—one can really argue that spiritualism does perform miracles.

This new image, however, is by no means universally accepted. The independent dealers throughout United States and Canada who are being squeezed out by Exxon's and Imperial's self-service stations do not accept it. Neither does the U.S. Federal Trade Commission, which in 1973 filed a complaint against Exxon and the other major oil companies for having "maintained and reinforced a noncompetitive market structure in the refining of crude oil" and having forced retail outlets to close and to "significantly curtail retail operations."[2] Thus Exxon is apparently continuing essentially the same practices which the U.S. Supreme Court so roundly condemned in its 1911 decisions.

What is more interesting is that even that staunch bulwark of conservatism, *The Edmonton Journal,* questions the new image of Exxon and its Imperial Oil satellite. In commenting on the Nova Scotia Power case the *Journal* of May 30, 1975 wrote:

> The federal government's planned sally into the petroleum industry via Petro-Can has been drawing impassioned objections from the private sector.
>
> But if ever there was offered a compelling argument against those objections and in favour of precisely such government intervention in the marketing end of the oil business, it has been offered by Imperial Oil.
>
> Imperial, it turns out, acting in connivance with its parent corporation, Exxon of the United States, managed to squirm through a loophole that enabled it to avoid paying Canadian taxes while hiking product prices for the Canadian consumers. . . .
>
> By weaseling its way out of paying taxes while at the same time hiking product prices, Imperial has given substantial moral ammunition to those subscribing to the 'corporate welfare bum' school of thought.
>
> It is difficult, however, to feel much sympathy for the oil company. The damage that has been done to its image of corporate citizenship has been done by its own hand.

The *Journal* is right. The tiger in the tank has not changed his stripes. But perhaps that is all to the good. Exxon may yet succeed in doing what no one else has been able to do—convert Alberta into a hot-bed of radicalism.

Exxon's power arises not only from its aggressive tactics but also from its relationships with other companies. John W. Wilson, formerly Chief of the Division of Economic Studies, U.S. Federal Power Commission, noted that "the key structural feature of the petroleum industry is that virtually all of its corporate entities are extensively tied together through a very large number of joint venture arrangements and other types of intercorporate links."[3]

Exxon has joint ventures with most of the major oil companies. Together with Standard Oil of California, Mobil and Texaco, it was, until very recently, the owner of Aramco in Saudi Arabia. This company is either now or will soon be owned entirely by the Saudi government. Exxon, Mobil and British Petroleum are joint owners of the Iraq Petroleum Company. Exxon, Gulf and Shell, until very recently, shared the output of the Mene Grande Oil Co. in Venezuela. That company has also been recently nationalized. Exxon, Shell, British Petroleum, Mobil, Compaigne Francaise Petrole, Texaco, Gulf, Standard Oil of California and a number of smaller American companies are all participants in the Iranian consortium, which was established after the CIA overthrew the Mossadegh government. Exxon is a partner with Sohio, British Petroleum, ARCO, Mobil, Union Oil, Phillips Petroleum and Amerada Hess in Prudhoe Bay and the Alyeska Pipeline. Many more examples of these joint-ventures could be cited.

These joint-ventures enable Exxon, the largest of the partners, to have substantial control over the oil supply in the non-communist world. In an interesting study of these joint-ventures, Prof. John M. Blair, formerly an economist for the U.S. Senate Antitrust Subcommittee, noted that between 1950 and 1972 the rate of growth of production in the OPEC countries was 9.55 per cent each year. This steady growth rate, Blair pointed out, did not come about accidentally but as a result of an international control mechanism developed by the partners in these joint-ventures. He described this control mechanism in detail. One interesting feature of it was that while production grew at a steady rate of 9.55 per cent in the OPEC countries as a whole, it grew at a faster rate in some individual countries and at a slower rate in other countries. By this method the oil companies were able to reward those countries which offered them more favourable conditions and to punish those which did not.[4]

The oil companies were able to mete out this differential treatment because they had a surplus of oil under their control. Thus, when Mossadegh nationalized British Petroleum in 1951, the oil companies were able to boycott that country and still expand production by 9.55 per cent by increasing output in Kuwait. Similarly, when Venezuela increased its income tax in 1959, the oil companies increased production in other countries at the expense of Venezuela.

In the so-called energy crisis of the last few years the oil companies lost, at least temporarily, control of this surplus. The OPEC countries were, for the first time, able to stop the companies from playing one country off against another. One of the chief aims of the oil companies and of the United States government is to create a new surplus of oil (and other forms of energy) so that the companies again will have the power to reward their friends and punish their enemies.

I mention this now because Imperial and the other multinational companies are emphasizing how much they are investing in Canada. These investments will give them greater control over Canadian resources. Assuming that they are able to create a new surplus, then they will be able to treat Canada just as they treated Iran and Venezuela. If Canada steps out of line, production will be cut back in Canada and increased elsewhere. This could have serious repercussions because these large investments by Imperial and other companies tend to lock the Canadian economy into the resource industries and thus make it more vulnerable to this type of pressure.

Exxon can act against Canada not only because of its joint-venture links but also because of its links through interlocking directorates.

Here is how Wilson describes some of these links:

> . . . the Chase Manhattan Bank has director interlocks with Exxon, Standard of Indiana, Shell, Marathon, Crown Central and Allied Chemical. The Chemical Bank has interlocks with the boards of Mobil, Marathon, Exxon, Standard of Indiana, Texaco, Allied, Champlin and Shell. First National City Bank has director affiliations with Exxon, Phillips, Champlin, Standard of California, Allied, Commonwealth Oil, Mobil and Monsanto. . . .[5]

These directorship ties are further reinforced by ownership links through the stockholdings of the Rockefeller interests. The Rockefeller interests control approximately 8.6 million shares of Exxon, making them the largest single interest group in the company. Exxon's 224 million shares are held by over 700,000 stockholders. The Rockefeller group controls about four per cent of these shares and through them has effective control over the company. The Rockefeller shares are either owned directly by members of the family or through foundations like the Rockefeller Foundation or the Rockefeller Brothers Fund, which the Rockefellers control, or are held in a fiduciary capacity by the trust

Exxon's Tankers: They also Serve...

THE CHAIRMAN: Do you own most of your tankers?
MR. MUNROE *(Comptroller, Exxon Corp.)*: Senator, I can put it in the record later but I think we own 50 per cent of our tanker coverage in our international business.
THE CHAIRMAN: Does it help to give this country a little more leverage in the worldwide oil picture which might be used to our advantage at some point if foreign countries decide they are going to be extremely unfair to the United States if the United States could demand that your company and others who have their own tankers simply not move the oil from countries that are acting against what appears to be in the best interests of this country?
MR. WRIGHT *(Chairman and Chief Executive of Exxon Co., USA)*: Of course, over the years we have had a tanker fleet of some magnitude and during a couple of world wars why, of course, our ownership of these tankers has been very important, and national interest considerations are very high in this.
THE CHAIRMAN: I know most of those fleets are not under U.S. flags. They are under flags of convenience, I believe.
MR. WRIGHT: True.
THE CHAIRMAN: Are they nevertheless subject to arrangements whereby they can be used to implement this country's foreign policy if they are required for that purpose?
MR. WRIGHT: We maintain control of their operation.
THE CHAIRMAN: So you maintain control of it and even though it is under a third country flag you would still be able to make your ships available to the United States if our interests required it.
MR. WRIGHT: We always have.
THE CHAIRMAN: It has always been that way and you would hope it would always be that way.
MR. WRIGHT: Yes, sir.

SOURCE: U.S. Senate, Committee on Finance, Hearings, *Profitability of Domestic Energy Company Operations,* Washington: U.S. Government Printing Office, Feb. 13-14, 1974, 180-1.

department of the Chase-Manhattan Bank, which the Rockefellers have controlled since the 1930's. In addition to being the largest single stockholder group in Exxon, the Rockefellers have substantial holdings in Mobil, Standard of Indiana and Standard of California. Nelson Rockefeller personally owns 308,000 shares of Exxon and 206,000 shares of Standard Oil of California. So much for the state of competition in the oil industry.

And, the Rockefeller and Exxon links extend far beyond the oil industry. Exxon has directorship links with such companies as Caterpillar tractors, IBM and General Motors, respectively the 36th, the 9th and the 2nd largest industrial corporations on *Fortune's* 500 list.[6]

The Rockefeller family controls the international Basic Economy Corp. (IBEC). It is run by Rodman Rockefeller, the son of Nelson. It operates in 30 countries, deals in housing, supermarkets, mutual funds, coffee, poultry and fish and has sales of over $300 million per year. The Chase-Manhattan Bank owns the largest single block of stock in United Air Lines, Northwest Airlines, ARCO, and National Airlines. It has substantial holdings in A. T. & T., IBM, Sperry Rand, International Paper, Motorola, Avon Products, Safeway and in that paragon of political purity, ITT. It also has considerable shares in the American Broadcasting Company, the Columbia Broadcasting System and the National Broadcasting Company and small shares in *The New York Times* and in *Time-Life*. Apparently the Rockefellers are still trying to improve their image.

Nelson Rockefeller personally has shares in the following companies: du Pont, General Electric, Merck, Eastman Kodak, Caterpillar Tractor, IBM, and Minnesota Mining and Manufacturing. The Rockefellers have indeed acquired quite an economic empire. John D.'s tactics have paid off handsomely.

But this economic empire could never have been built and certainly never maintained if it were not buttressed by political links.

The political links are of two types: general and specific. The general links are those flowing from the overall policies of the U.S. government towards business while the specific links are those of the Rockefeller-Exxon axis with governmental agencies. I shall first discuss these general links.

The general links arose from the growth of the American Empire. After slaughtering the Indians from "sea to shining sea", the Americans began to look for new lands to conquer. Their tactics were, however, to some extent different from those of the European imperial powers. With the exception of Puerto Rico, Hawaii, the Philippine Islands, the Virgin Islands and the Panama Canal, the Americans generally did not acquire colonies as such. Instead, and in this they were much more sophisticated

Exxon: What's in a Name?

SENATOR CHURCH *(Chairman, U.S. Senate Subcommittee on Multinational Corporations):* What was the cost to the company of changing the name to Exxon?

MR. PIERCY *(Sr. Vice-President and Director, Exxon Corp.):* My recollection is that these costs involved substituting something for something else in advertising, so they are not really added costs.

SENATOR CHURCH: Signs?

MR. PIERCY: Well, more than that.

SENATOR CHURCH: We saw the signs change.

MR. PIERCY: Well, your media advertising, if you are doing a name change, you spend your money on the name change and not on something else. It is a matter of keeping your name before the public.

SENATOR CHURCH: Yes.

MR. PIERCY: So they are not necessarily true added costs. It was, and I am just a little hazy, but it is something approaching $100 million before these offsetting credits.

SOURCE: U.S. Senate, Subcommittee on Multinational Corporations of the Committee on Foreign Relations, *Multinational Corporations and U.S. Foreign Policy,* Washington: U.S. Government Printing Office, Part 5, Feb. 1, 1974, 204.

than the older imperial powers, they acquired effective control through economic penetration of other areas, especially Latin America and Canada.

As part of this policy the U.S. government has consistently supported the activities of U.S. companies abroad, even at times when it was prosecuting these very same companies for their behaviour at home. The oil companies, especially, received the support of the U.S. government.

This followed from the policy of oil imperialism adopted by the U.S. government at the end of World War I. During that War the U.S. became aware of the importance of oil to the armed forces. The U.S. furnished over 80 per cent of the Allies' petroleum needs. In order to do so and to supply its own forces as well, it imported oil from Mexico and instituted a programme of "gasless Sundays" in which people voluntarily refrained from using their automobiles except for "errands of necessity."

These wartime shortages gave rise to fears that the U.S. might not have sufficient oil resources to see it through future emergencies. In fact there were dire predictions that the U.S. would run out of oil in five to ten years, so similar to predictions we hear today.

Disturbed by the prospect that the U.S. might not have sufficient reserves for a future war, Mark Requa, General Director, United States Fuel Administration, advocated a policy of encouraging "in every way possible . . . the acquisition by our nationals of reserves in foreign lands."[7] In pursuance of this policy, the U.S. used both economic incentives, such as tax concessions, and diplomatic pressure to ensure that the U.S. firms had access to reserves abroad.

After World War I, for instance, the U.S. almost went to war with Britain over the latter's refusal to allow American oil companies in Iraq. As a result of U.S. diplomatic pressure the British reluctantly agreed to allow in two American companies. They were Mobil and Exxon.

Similar diplomatic pressure was used against Holland when in 1921, the Dutch government granted an exclusive concession in the Djambi region of Indonesia to a company jointly owned by the Dutch government and Royal Dutch Shell. Exxon and another U.S. company, Sinclair, wanted access to this field, regarded as the richest in the Netherlands East Indies. Irked by the refusal of the Dutch government to let them in, they came to the State Department for aid.

In a language very similar to that used by American ambassadors in Ottawa today, the State Department complained that "a monopoly of such far-reaching importance in the development of oil is about to be bestowed upon a company in which foreign capital other than American is so largely interested." The Department then pointed out that "American capital stands ready to assist in the development of the Djambi fields and other oil deposits in the Netherlands Indies."[8] After several years of negotiation, the United States and Holland reached an agreement whereby Exxon, but not Sinclair, received a concession in Indonesia.

The United States government also played a direct role in opening up Iran to American oil companies. Until 1951, when Mossadegh nationalized the oil industry, Iran had been the private preserve of British-Petroleum. After the overthrow of Mossadegh, the State Department negotiated an agreement which gave American oil companies a 40 per cent share of the Consortium. The Department then invited Exxon and four other U.S. Companies (Gulf, Texaco, Mobil and Standard Oil of California) to participate in the Consortium even though they were then being sued by the Department of Justice for antitrust violations in the international field.

There is little question that the U.S. is still encouraging American companies to explore abroad in order to add to the *U.S. strategic oil reserve.* However this exploration will probably now be confined to those areas which are considered "safe", i.e. securely in the U.S. orbit. The implications of this policy are very significant for Canada, especially when Exxon's Canadian satellite, Imperial, proudly publicizes the amounts it is investing in exploration in Canada.

The problem with the U.S. strategy is that areas which are considered "safe" today may not be "safe" tomorrow. Exxon, like the other oil companies, is aware of this possibility and therefore has taken steps to keep "safe" areas "safe".

One area which has caused particular concern to Exxon is Italy. Exxon expressed its concern by contributing through its Italian satellite, Esso Italiana, between $46 and $49 million to Italian political parties between the years 1963 and 1972. The bureaucracy in Esso Italiana was so used to making these contributions that one employee routinely made out a voucher of $86,000 to the Italian Communist Party. This error was apparently rectified. The contribution did not reach the Party.

When asked by the U.S. Senate Subcommittee on Multinational Corporations why Exxon made these contributions, Exxon executive Archie L. Monrow replied, in good image-making terms, "to further democracy." Exxon is also apparently interested in "furthering democracy" in Canada. Its Canadian satellite has been making political contributions of about $234,000 a year to Canadian political parties, more than any other oil company in Canada. Nevertheless this is chicken-feed compared to the $5 million a year spent in Italy. Canadian politicians should take a lesson from this. It just doesn't pay to be taken for granted.

But more important than Exxon's links with political parties and presumably governments in host countries are its special links with the

U.S. government. Especially since the end of World War II, Exxon and the Rockefellers have played a leading role in formulating U.S. foreign policy. John Foster Dulles, architect of the Cold War, came to the State Department from the law offices of Sullivan and Cromwell. Sullivan and Cromwell were then the attorneys for Exxon. Dean Rusk, whose brainchild was the Vietnam War, came to the State Department from the Rockefeller Foundation. Henry Kissinger also journeyed to the State Department from Nelson Rockefeller's office. Nelson even gave Henry a parting gift of $50,000 to enable Henry to perform his public duties more efficiently. Probably no other single interest has had as much influence on U.S. foreign policy as the Exxon-Rockefeller group. Is there any wonder that U.S. policy makers view Exxon and U.S. interests as one and the same?

We can now see how the political and economic power of Exxon are really two sides of the same coin. One could not exist without the other. Exxon is a microcosm of modern capitalism, in general, and of American imperialism, in particular. This is both its strength and its weakness.

Historically, all empires have come to their ends. The American Empire is no exception. It too is beginning to decline. As events in Vietnam and Angola have shown, America is no longer invincible. Given the new world situation, Exxon and the other oil companies may no longer be able to wield the power they held in the past. And given the new world situation, Canadians should ask themselves whether it would be in their interests to continue to allow this company to exploit their resources.

---

1 U.S. Supreme Court, *221 U.S. 1*, 1911.

2 U.S. Federal Trade Commission, *Docket No. 8934*.

3 *The Journal of Economic Issues*, June 1975, 326.

4 *Ibid.*, 297-318.

5 *Ibid.*, 327.

6 *Fortune*, May, 1975.

7 U.S. Fuel Administration, *Final Report of the United States Fuel Administration, 1917-1919*, Washington: U.S. Government Printing Office, 1921, 272.

8 U.S. Congress, Senate, Special Committee Investigating Petroleum Resources, *Diplomatic Protection of American Petroleum Interests in Mesopotamia, Netherlands, East Indies and Mexico. A Study Prepared for the Special Committee Investigating Petroleum Resources by Henry S. Fraser*, Document No. 43, 79th Congress, 1st Session, 1945, 41-2.

Exxon Owns the Fruits of Imperial's Research

Under the Standard Research Agreement between Exxon Corporation and it subsidiaries, including Imperial Oil, Exxon owns all of the technology developed by Imperial's employees and research groups in Canada. When an Imperial researcher develops a new idea, it becomes the property of Exxon. This has ominous implications for Canada in terms of technological backwardness. It has special implications for joint ventures like Syncrude in which three governments are partners of Imperial. Will Syncrude's oil sands technology end up being owned by Exxon? Quite possibly.

J.A. Armstrong, Imperial's Chief Executive Officer, was quoted in the *Imperial Oil Review* as stating:

> Syncrude will certainly provide unique technical expertise to some of our international affiliates that may need it. Another example [of Imperial's contribution to the overall research of Exxon] is our basic research into lubricating-oil technology. Imperial is doing the work in this field for the Exxon organization. As in all mutual research, the work is underwritten by the participants of our research agreement and the results are distributed in turn.

In March, 1976 Exxon became concerned that subsidiaries were acting as though they controlled their own research. The parent corporation moved to bring its foreign branches, including Imperial, into line. The following two letters were sent to Imperial from Exxon.

March 4, 1976

Imperial Oil Limited
P.O. Box 4029, Terminal A
Toronto, Ontario, Canada

Dear Sirs:

From time to time we have become aware of instances in which an affiliate having an SRA [Standard Research Agreement] with us, without prior consultation with us, has made commitments to third parties involving patent rights or technical information. In certain cases commitments made to third parties have been inconsistent with prior contractual obligations to us with the result that the commitments to the third parties could not be fully met.

The enclosed letter from Dr. N.V. Hakala to the SRA affiliates discusses the respective rights of Exxon Research and Engineering Company and the affiliates in patent rights and technical information under the SRA. It suggests that we be consulted before any commitments are made to third parties involving patent rights or technical information and recommends that we be consulted at an early stage in the negotiations so as to avoid possible embarrassment arising from making tentative commitments which cannot be honored.

We would greatly appreciate it if you would circulate copies of Dr. Hakala's letter to the departments in your organization likely to make arrangements with third parties which could involve patent rights or technical information. We would also appreciate it if you would refer any doubtful cases to us in the event any question should arise as to whether SRA rights and obligations are involved.

Very truly yours,
W.V. Smith

March 4, 1976

TO ALL SRA AFFILIATES:

From time to time we observe evidence of possible misunderstandings by our affiliates of their rights and obligations under the SRA in respect of patents and technical information. In the interest of minimizing possible misunderstandings we thought it might be well for us to furnish you with a brief statement of the rights and obligations of the parties under the SRA respecting patents and technical information.

Patent rights in the field of the SRA are vested in Exxon Research and Engineering Company (ER&E) without regard to whether the work out of which the invention arose was performed by ER&E itself or by an affiliate, and this is so whether or not the cost of the research in question is borne by ER&E. Each affiliate has a patent license under the SRA to carry out petroleum or chemical operations for its own account. It has no power to grant patent rights of any kind to third parties. In appropriate situations, ER&E will license its patent rights for the common benefit of the SRA parties.

Somewhat similarly, the affiliates are licensed to use all technical information in the field of the SRA in petroleum or chemical operations for their own accounts. Affiliates are not permitted by the terms of the SRA to disclose or license technical information of a confidential nature to any third party, with very minor exceptions in respect of disclosure to customers and contractors. ER&E alone has broad rights to disclose and license such confidential technical information, which rights it exercises for the common benefit of the SRA parties.

Occasionally affiliates have made actual or tentative arrangements with third parties, the implementation of which would involve a violation of the principles mentioned above. If an affiliate is considering any arrangement with a third party involving rights in confidential technical information or patents in the field of the SRA, whether the information or patents are in existence or thereafter to be developed, ER&E ought to be consulted with respect to the arrangement. Very possibly, ER&E ought to be the party to the agreement, depending on its content. If the practice of early discussion with ER&E is not followed, it may become embarrassing if, in the exercise of its responsibilities, ER&E does not find it feasible to honor tentative commitments with respect to technology and patent rights which an affiliate may have made in the mistaken belief that it was free to do so.

In any case in which an affiliate requires assistance in connection with any proposed arrangement, ER&E will be glad to render whatever help it can and will endeavor to facilitate the ends of the affiliate so far as it can do so consistent with the proper exercise of ER&E's own responsibilities. We would greatly appreciate being kept informed of all proposed arrangements in the technology field at a very early stage in their consideration.

Very truly yours,
N.V. Hakala

Esso

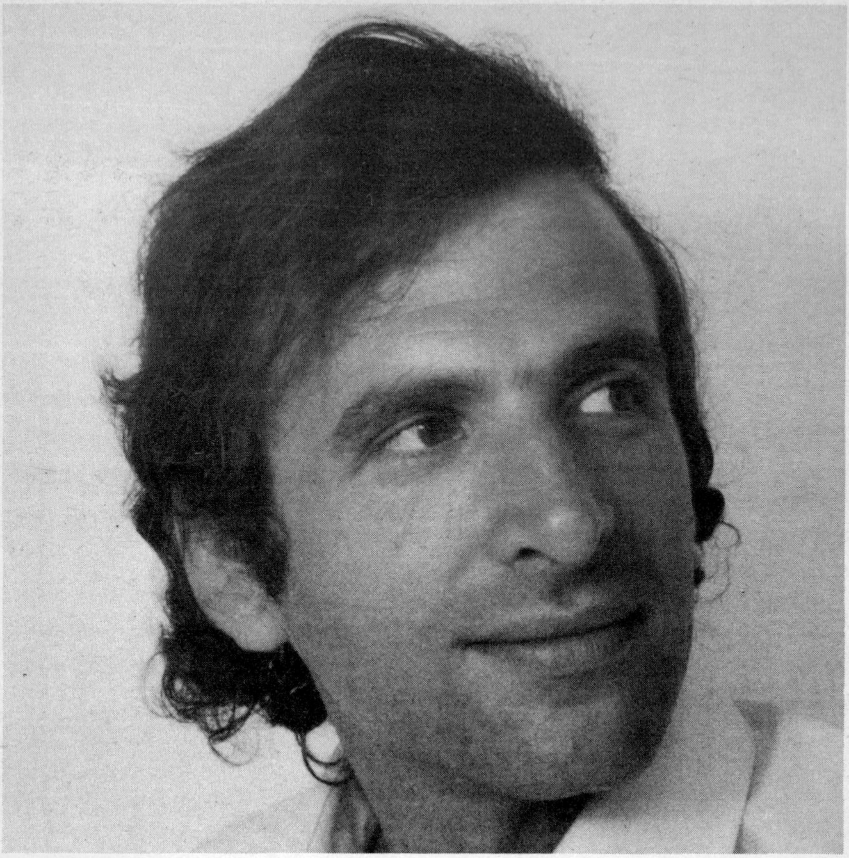

JAMES LAXER

James Laxer has been a leading critic of the petroleum industry since 1970. He is the author of *The Energy Poker Game* (1970) and *Canada's Energy Crisis* (1974). An Associate Professor of Political Science, James Laxer is Chairman of the Department of Political Science at Atkinson College, York University. He is on the National Executive of the Public Petroleum Association.

2

James Laxer

Always Look to Imperial for the Best: Imperial Oil and Canadian Energy Policy

On April 26, 1976 J.A. Armstrong, President and Chief Executive Officer of Imperial Oil, took the occasion of his company's annual meeting to answer the charge that, in his words, "the oil companies form part of a foreign conspiracy that is intent, in some undefined manner, on defrauding Canadians of an important natural resource."

Armstrong told the shareholders:

> There is no doubt that, so far as Imperial is concerned, the accusation stems from the fact that approximately 70 per cent of the company's shares are owned by Exxon Corporation, giving rise to the suspicion that a substantial portion of this country's energy is effectively controlled by a foreign organization, which is at liberty to export our resources at will.
>
> Governments, not the oil industry, decide on the disposition of this country's energy resources and this, of course, is as it should be. The experience of the past two years, when major increases in crude oil prices and royalty rates brought little financial benefit to the producer, certainly demonstrates who calls the tune. All we do is dance.

Armstrong then went on to discuss Imperial's relationship with its chief dancing partner, Exxon Corporation:

> So far as your company is concerned, anyone who believes that Imperial is 'manipulated' by Exxon in that company's interests is unaware of the nature of the Imperial-Exxon relationship.[1]

In these capsule comments, Armstrong presented the view that Imperial likes Canadians to have of its operations. The image presented is that of hard working, competent, Canadian-managed company doing its best to supply Canadian consumers and industries with oil and gas; a company thoroughly committed to the national interest and so closely regulated by government that it has little room to manoeuver on its own.

Jack Armstrong is the personification of Imperial's carefully groomed image. A Canadian himself, a native of Dauphin Manitoba, Armstrong was described by *The Globe and Mail* Report on Business as a man possessed of a confident and yet restrained corporate image.[2] A member of Imperial's board of directors since 1961 and President of the company since July 1970, Armstrong became Chief Executive Officer when W.O. Twaits, his legendary predecessor, retired in 1974. Armstrong serves both as a director of the company and as an officer along with nine other men. The ten received an aggregate salary of $1,263,000 in 1975 for an average of $126,000.[3]

Directors and officers of Imperial Oil are noted for their fierce loyalty to the company. Many of them have a moderate to large stake in the company themselves in terms of ownership of shares. As of March 1976, Armstrong owned 18,405 Class B Imperial shares.[4] At $23 per share, his holdings in Imperial stock were worth $425,000.

But to put Armstrong's position in perspective, his $425,000 holding should be seen alongside that of the company's largest shareholder, Exxon Corporation of New York. Exxon owns shares in Imperial worth just over one billion dollars of the $1.4 billion in shareholders' equity.

Despite his professions of Canadian control to the recent meeting of Imperial's shareholders, Armstrong knows where the ultimate power lies. Interviewing him earlier this year, Raoul Engel of Global Television put the tough question to Armstrong: "But Mr. Armstrong, if it came down to the crunch with Exxon, if it got down to out-and-out disagreement, to bare knuckles, what then?"

Armstrong replied:

> If it got down to bare knuckles? Exxon would either go along with Imperial's executive committee and with me as Chief Executive Officer, or they could get rid of us. It's that simple.[5]

Although Imperial will celebrate its centenary in 1980, its origins go back exactly 100 years to 1876 when several small refiners came together to form the London Refining Company. Four years later, the London Refining Company along with 15 independent producers and refiners united in the formation of the Imperial Oil Company Limited.[6] Appropriately enough, the new venture was undertaken to establish an organization strong enough to fight off competition from across the border, from the Standard Oil Company, the creation of John D. Rockefeller.

For some years Imperial and Standard competed for markets in Canada. In February 1899, however, Standard Oil acquired control of Imperial and Standard's Canadian affiliates were merged with the formerly Canadian company under the Imperial name.[7] Imperial was involved in the early development of Alberta's oil field in the Turner Valley in the 1920's and later made the famous discovery at Leduc, Alberta in 1947 that ushered in the modern age of Canadian petroleum development.

Best known to Canadians through its 5900 franchised service stations, Imperial Oil today is by no means a simple producer of oil and gas. It is a giant corporate institution, whose activities are carried out through five divisions. The divisions are: 1. oil and gas production, 2. chemicals and building materials, 3. transportation, 4. other investments, 5. refining and marketing. For competitive reasons, Imperial Oil does not break down its earnings division by division in its report to shareholders. However, a Calgary based investment consulting firm has produced an analysis of each of the operating divisions of the company.

The oil and gas producing division is by far the most important. At the end of 1974, the division controlled 1.2 billion barrels of proven oil reserves. 94 per cent of these are located in Alberta where Imperial's oil reserves are highly concentrated, 55 per cent of them occuring in five large, prolific fields. In 1975 Imperial produced an average of 265 thousand barrels of crude oil and natural gas liquids daily, over one-eighth of a total average Canadian production of 1,780 thousand barrels. This was down from the company's peak production of 345 thousand barrels daily in 1973. The decline in production has been due to cutbacks in exports to the U.S. and decline in petroleum sales in Canada in the past year. In 1975, Imperial refined an average of 395 thousand barrels of crude oil daily and sold petroleum products amounting to an average of 418 thousand barrels a day, each of these figures representing more than 20 per cent of the national total.[8]

While Imperial is the nation's major crude oil producer, it is a less significant natural gas producer, averaging sales of 412 million cubic feet daily in 1975, an amount that constitutes about 6.2 per cent of industry sales. Of Imperial's after tax profit in 1975 of $250 million, $154 million is estimated to have been generated from the company's oil and gas production division.[9]

Imperial's Chemical and Building Materials division operates major units that produce petrochemicals in Ontario and Alberta and various kinds of building materials through Imperial's subsidiary, Building Products of Canada. This division is estimated to have contributed about

Like to guess how much profit Imperial Oil earns on each gallon of petroleum products it makes and markets?

If you guess one to two cents a gallon, you're close.

Actually, in 1975, the figure was less than one cent a gallon.

That's right – after paying income tax, Imperial earned 9/10ths of one cent on each gallon of petroleum products it made and marketed during 1975. Products like gasoline, heating and fuel oils, jet fuels, greases, oils and many more.

In other words, if Imperial reduced the prices of these products by an amount equal to their after-tax profit, their average price would be lowered by slightly less than one cent a gallon.

Imperial has other sources of earnings – from the production of crude oil and natural gas, from the manufacture and sale of chemicals, from other sources.

And the bulk of Imperial's earnings is reinvested in the big job of finding and developing new petroleum resources for Canada, and in providing facilities to meet the growing demand for products.

We hope you'll remember this message the next time you hear about...or read about...or even think about...the price of petroleum products.

Esso Imperial Oil Limited.
Canada's leading supplier of energy.

MACLEAN'S APRIL 5 1976

$20 million to the company's 1975 profits, or about one-eighth as much as the oil and gas production division contributed.[10]

The Company's Transportation division includes ownership of product pipelines, crude oil gathering systems, terminals for bulk products, tankers and barges. The division is also involved in land development through an Imperial subsidiary, Devon Estates Ltd. The net earnings of this division are estimated at $15 million for 1975.[11]

The fourth category, other investments, includes a 32.8 per cent equity in Interprovincial Pipe Line Ltd., the transportation system for shipping western Canadian crude oil to eastern Canadian and American markets. Profits in this category were estimated at $17 million for 1975.[12]

Finally, Imperial's Refining and Marketing division is estimated to have earned a profit of $41 million in 1975.[13] This division has been subject to brisk competition with other oil companies because Canadian oil refining capacity is now overextended relative to current demand levels. This division gives the company the opportunity to engage in some highly questionable advertising in Canadian magazines. For example, an Imperial ad appearing in *Maclean's* on April 5, 1976 informed readers that Imperial's profit was less than one cent on each gallon of petroleum products it makes and markets. This refers to the operations of the refining and marketing division of the company. In small print in the ad there is a further statement: "Imperial has other sources of earnings—from the production of crude oil and natural gas, from the manufacture and sale of chemicals, from other sources."[14]

As we have already seen only about 20 per cent of Imperial's profits come from the division of the company on which the claim of the ad is based. If profits from the oil and gas production division and the transportation division of the company are included in calculating profit per gallon on petroleum products, the proper figure would be much higher than the company states in its advertisement.

Taken together, Imperial's five divisions generated a total revenue for the company of $4,110 million in 1975, which means that over three per cent of all expenditures in the Canadian economy that year were made to this single corporation.

In the speech to Imperial shareholders cited above Jack Armstrong stated that his company is thoroughly regulated and that the basic decisions concerning oil and gas policy are made by governments. Having looked at the size and structure of the Imperial organization, let us examine the ways in which it has used its power to shape Canadian economic and political decisions.

As everyone knows from Imperial's messages on *Hockey Night in Canada*:

Each year Imperial spends hundreds of millions of dollars on the big, tough, expensive job of developing petroleum supply, and if the company is to be successful in helping Canada reduce its dependence on foreign sources, the company will have to put even more money to work.[15]

Despite the message, if Imperial has had a fundamental historic policy that it has pressured Canadian governments to accept, it has been the pressure to increase the export of Canadian oil and natural gas to the United States. For all Jack Armstrong's talk about Imperial's commitment to Canada's interest, his company was the prime driving force behind the export of Canada's cheap oil and gas reserves in the prairies during the 1960's and the early 1970's.

As early as 1960 Imperial included the following statement in its annual report:

Our view is that opportunities are open for Canada to move ahead in the '60's, providing we do not afflict ourselves with a costly and emotional economic nationalism.[16]

The 1963 annual report spoke glowingly of "the continued evolution of the now well established Canada/U.S. pattern of energy movements so essential to full utilization of western Canadian oil and gas resources."[17]

The annual report for 1969 appeared after the publication of the Shultz Report in the United States, the task force report to the Nixon administration on American imports of oil and consequent problems of American national security. Both the majority and minority versions of the report called on the U.S. government to evolve a continental energy resources deal with Canada. Imperial's report echoed the possibilities:

. . . other major factors will influence the place of Canadian crude in the North American supply pattern. One is the outcome of the current comprehensive U.S. government examination of its oil import policy: both the majority and the minority reports of the cabinet committee appointed to conduct the imports study have stressed the need for security of oil supply and the special place of Canadian oil in this framework.

Currently Canadian wells have the capacity to supply a larger share of the North American market than they are now serving. And there is reason to look forward to discoveries of major new reserves in Canada.[18]

In July 1970 when Armstrong became President of Imperial Oil he told *The Globe and Mail* that his company was in favour of the idea of pooling North American energy sources in a continental policy. He said he felt certain that the United States would need increasing quantities of Canadian oil, but that a Canadian commitment to a continental approach was needed so that the industry could begin to formulate long-term planning. Armstrong appeared to be unimpressed by the debate of the day on whether a continental policy was a good thing. He had already made up his mind. He stated that Imperial was already proceeding with its exploration programme in the belief that the two governments would eventually harmonize their policies on resource development and use.[19]

Again in 1971, the annual report urged:

> The industry must be able to rely on exports to the United States if it is to undertake the financial commitments necessary to unlock the petroleum potential in the north and the offshore areas.[20]

And finally, in 1972, the year before the international price of oil quadrupled and changed Imperial for all time, one last strident defence of exports in the annual report:

> In the current debate, the export of Canada's energy resources is being questioned; in effect, we are being urged to 'bank' our petroleum resources. Our present energy reserves, using present technology, are sufficient for our requirements for several hundred years.
>
> Export markets will not wait on our convenience. Hydrocarbons can be secured from other geographic sources, and technology leading to the development of other energy forms is being pursued vigorously. In other words, once lost, export markets cannot easily be regained—assuming they can be regained at all—and their loss would be a genuine economic setback for Canada.[21]

Throughout the sixties and the early seventies, Imperial kept up its pressure for exports and its insistence that the country had the oil supplies to meet domestic and foreign demand. In 1969 the Geological Reserves Committee of the Canadian Petroleum Association, the voice of the private petroleum companies, the largest of which was Imperial, issued a report indicating that Canada had undiscovered reserves of over 100 billion barrels of oil.

In June 1973, just a few months before the quadrupling of world oil prices, the long heralded report of the Department of Energy, Mines

The Change In Oil Producibility Estimates 1973—1976

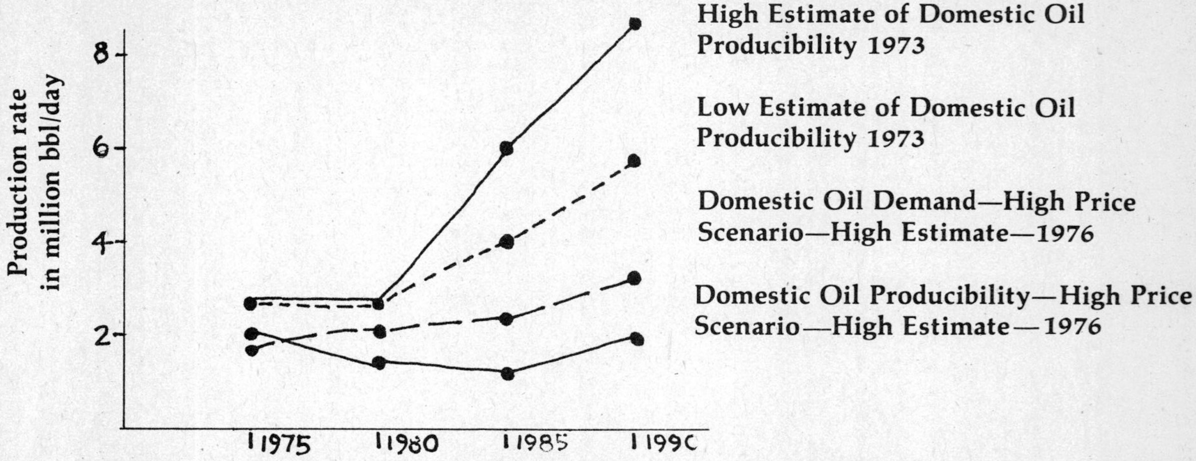

Source: Canada, Dept. of Energy, Mines and Resources, *An Energy Policy for Canada—Phase I,* Ottawa: Dept. of Energy, Mines and Resources, 1973, v. 1, p. 102-3

Canada, Dept. of Energy, Mines and Resources, *An Energy Strategy for Canada: Policies for Self-Reliance,* Ottawa: Minister of Supply and Services Canada, 1976, pp. 55 & 65.

and Resources entitled *An Energy Policy for Canada* was published. Coming at the end of a decade and a half of sweet words of encouragement from the petroleum industry led by Imperial, the report confidently predicted that Canada could easily meet its oil needs to at least the year 2050 and for prices below $7 a barrel.[22] For the period to the end of the century the report ventured a pessimistic prediction and an optimistic prediction: the pessimistic prediction was that from the early seventies to the year 2000, oil output would increase from about two million barrels a day to six million barrels a day; the optimistic prediction suggested that Canada might reach 10 million barrels a day.[23]

The report came only a few months after the Alberta government had received a report from Walter J. Levy, the renowned New York oil consultant, who had predicted that ventures in the oil sands would be profitable at a crude price of $5 a barrel.[24]

Then came two events that caused the old nostalgic world of petroleum to vanish: the freezing of the price of Canadian crude oil at $3.80 a barrel in September 1973 and the quadrupling of the world oil price from $3.00 a barrel to $11.00 a barrel in December 1973.[25] Suddenly, the press releases coming out of Calgary and Toronto oil offices were different. They carried a new refrain: a warning. The warning is now familiar to all Canadians. It goes like this: if the price of domestic oil is not allowed to rise to the world price, Canada will face oil shortages by the end of this decade or the beginning of the next. If the petroleum industry is not given adequate incentive to explore for new reserves of petroleum, the nation could face a crisis of supply soon.

In the spring of 1974, the federal government raised the price of domestic crude oil to $6.50 a barrel. Several months later the National Energy Board held hearings on the question of Canadian oil exports. The major petroleum companies all made estimates of how much oil Canada could expect to produce to the year 1990. Imperial Oil, the industry leader, appropriately enough, submitted three estimates of its own. Stunningly, unanimously, the estimates of the petroleum industry were all markedly more pessimistic than the most pessimistic estimates on which the June 1973 report had been based. In October 1974 the NEB produced its report. The report concluded that by 1982 Canada would be unable to meet it oil needs west of the Ottawa Valley with domestic crude.[26] The eighty year surplus had become a seven and a half year surplus in the space of 16 months.

Now we have a newer report, issued on April 29, 1976 under the auspices of Energy Minister Alastair Gillespie. The latest report provides a stark contrast with its predecessor, the 1973 report. It says that the nation has 60 per cent less petroleum than was thought three years ago.

It predicts that even with the world price of oil for domestic crude (about $13.00 a barrel) Canada will have an annual deficit in its oil trade of $4.5 billion by 1985.[27]

No doubt concerned that the long suffering public might be having problems believing the latest report, the authors offered a terse three sentence explanation for the difference in estimates between the 1973 report and the 1976 offering. The explanation was as follows:

> First, the 1973 figures were based on a less sophisticated method using a geological analysis of all the sedimentary basins in Canada and comparing these with other petroleum-producing basins throughout the world. Second, significantly more information is now available from exploratory drilling, and it must be recalled that over the past two years drilling results in the frontier area have been very disappointing. Finally, these estimates are not strictly comparable because the areas included are not the same: the 1973 estimates included for example, the continental slope and rise, whereas the latest estimates do not.[28]

However reassuring it may be that the Geological Survey of Canada is now using more sophisticated methods of analysis and however heartened Canadians may be by the news that certain sedimentary basins included in the last report have been dropped from this one, it is difficult to avoid the conclusion that the decisive explanation is the one concerning drilling results over the period since 1973. Once again, it is the information provided by the petroleum companies that is the critical variable.

One of the report's conclusions implies that this indeed is the critical variable. The report states:

> To plan effectively it is necessary to know what resources are available. One of the most serious problems facing the federal government in its efforts to elaborate energy policies that are appropriate is the substantial degree of uncertainty that remains with regard to ultimately recoverable resources in the frontier areas of Canada. In the Beaufort Sea, and the offshore areas of the Arctic Islands and eastern Canada, large geological structures which may contain substantial reserves of hydrocarbons have been identified. It is difficult to develop appropriate energy policies without knowing whether these structures are full or empty, and this knowledge can be obtained only by drilling.

The report continues:

> The Government of Canada will introduce legislation providing the responsible Minister with the authority to require companies holding rights to such structures to have exploratory drilling undertaken within a reasonable time

period. . . . It will make possible, over a period of time, the receipt of information that is necessary to plan our energy future in the most beneficial way.[29]

Clearly the authors of the report fear that what subcommittees of the U.S. Congress have proven in a number of instances to be the case with the U.S. domestic petroleum industry, may be true for the industry in Canada. They fear that the industry is speculating on leaseholds, holding off drilling until the price goes up, until the government is panicked into guaranteeing the companies a higher rate of return.

There is another oddly disturbing feature about the latest energy report. Despite its dozens of tables and charts, the report fails to provide primary data on cost estimates for the costs of exploration and production of oil and gas in the various regions of Canada under scrutiny. The report provides information derived from these primary inputs but does not provide the primary data. All we see are the conclusions which may or may not be valid depending on the strength of the missing primary information on which the conclusions are based.

What the report is really about is an assessment of what level of returns will entice the companies to go out and do the "big, tough, expensive job" of developing new Canadian petroleum supplies. The theory that underlies the federal government's high price is this: more cash flow to the companies results in more exploration; more exploration will make accessible whatever reserves actually remain.

As W.J. Young, Senior Vice-President and Director of Imperial Oil, said in a speech in Winnipeg in March 1976:

> The oil industry's position can be very simply stated. It is not receiving a fair return on highly productive past investments. This impedes its ability to find and develop new sources of domestic petroleum . . . Its cash flow is lower than it should be.[30]

The great complaint of the companies has been that they have lacked the cash flow to do the job. Increase our cash flow they have said and the job will be done.

Using Imperial Oil as a case study let us examine the record to discover whether the theory that greater cash flow leads to greater exploration is, in fact, true.

We will examine Imperial Oil's performance from 1972 through 1975. The figures that follow will compare 1975 levels with the base year of 1972, the last year before the international oil crisis occurred.

In 1975 Imperial sold an average of 418 thousand barrels of petroleum

Do Higher Profits Lead to More Exploration? The Record of the Major Petroleum Companies in Canada: 1972 & 1975

	Revenue		Net Income		Oil Sales		Gas Sales		Exploration Expenditure		Net Income as % of shareholders' equity	
	'72	'75	'72	'75	'72	'75	'72	'75	'72	'75	'72	'75
Imperial	2104	4110	151	250	417	418	425	412	74	74	14.1%	16.4%
		up 95.3%		up 65.6%		up .2%		down 3.0%	no change			
Gulf	874.4	1729.9	66.4	176.6	245	265	499	470	28	54	9.7%	18.3%
		up 97.8%		up 166.0%		up 8.2%		down 5.8%		up 92.9%		
Shell	868	1868.4	69.2	144.8	251	270	602	621	10.2	23.5	9.0%	14.0%
		up 115.3%		up 109.2%		up 7.6%		up 3.2%		up 115.6%		
Texaco (Canada) Ltd.	463.4	878.4	42.4	51.1	176.4	181.7			5.0	5.1	16.1%	13.9%
		up 89.6%		up 20.5%		up 3.0%				up 2.0%		
BP	235.0	493.1	13.0	30.5	94.7	104.3	97	101	10.2	29.3	5.5%	9.8%
		up 109.5%		up 134.6%		up 10.1%		up 4.1%		up 187.3%		
Total	4544.8	9079.8	342.0	653.0	1184.1	1239.0	1623	1604	127.4	185.9		
		up 99.8%		up 90.9%		up 4.6%		down 1.2%		up 45.9%		

Dollar figures in millions
oil sales in thousands of barrels per day
natural gas sales in millions of cubic feet per day

Prepared by the Public Petroleum Association of Canada

products a day, a tiny increase of .2 per cent over the 417 thousand barrel a day figure for 1972.

In 1975 Imperial sold an average of 412 million cubic feet of natural gas daily, a decrease of 3 per cent from 425 million cubic feet in 1972.

In 1975 Imperial's gross production of crude oil and natural gas liquids totaled 265 thousand barrels a day, an increase of one per cent from 1972's production of 262 thousand barrels.

As can be seen from these figures, Imperial's production levels were virtually indentical in 1972 and 1975.

What about revenues?

In 1975 Imperial's operating revenue was $4,047 million, an increase of 98 per cent from the 1972 revenue of $2,104 million.

In 1975 Imperial's revenue from the sales of petroleum products was $2073 million, up 89.5 per cent from the 1972 figure.

Revenue for crude oil and natural gas sales in 1975 was $1,542 million, up 121 per cent from 1972.

Profits after tax were $250 million in 1975, up 65.6 per cent from 1972. Working capital from operations was $414 million in 1975, up 69.7 per cent from 1972.

And cash flow in 1975 (the critical catagory) was $464 million in 1975, up 53.6 per cent from 1972.

What about exploration for new sources of oil?

In 1975 Imperial spent $74 million on exploration for new oil, exactly the same amount it spent in 1972.

In 1975 Imperial drilled 6 new exploratory wells, down from 20 exploratory wells in 1972 and incomparably lower than the 90 new wells it drilled in 1968 when cash flow was just over one-third of the 1975 level.[31]

Imperial's commercial messages about the company's gigantic exploration disguise the fact that despite a large increase in revenues, profits and cash flow, the company has failed to step up its exploration effort. Even the amount of exploration actually done is carried out under the prodding of very generous tax incentives. Imperial can write off 100 per cent of its exploration expenses against its corporation tax and can write off an additional 33 per cent as an earned depletion. The tax incentive means that each exploration dollar costs the company only 39 cents. The other 61 cents is a foregone tax, a form of public investment. (Despite this, when Imperial informs the public about its rate of profit on the production of crude oil, it provides figures that do not take into account the effect of tax incentives for exploration and development. Imperial claims it makes an after tax profit of $1.69 a barrel on "old" Alberta crown oil. Once the tax incentives are added in, a truer figure would be

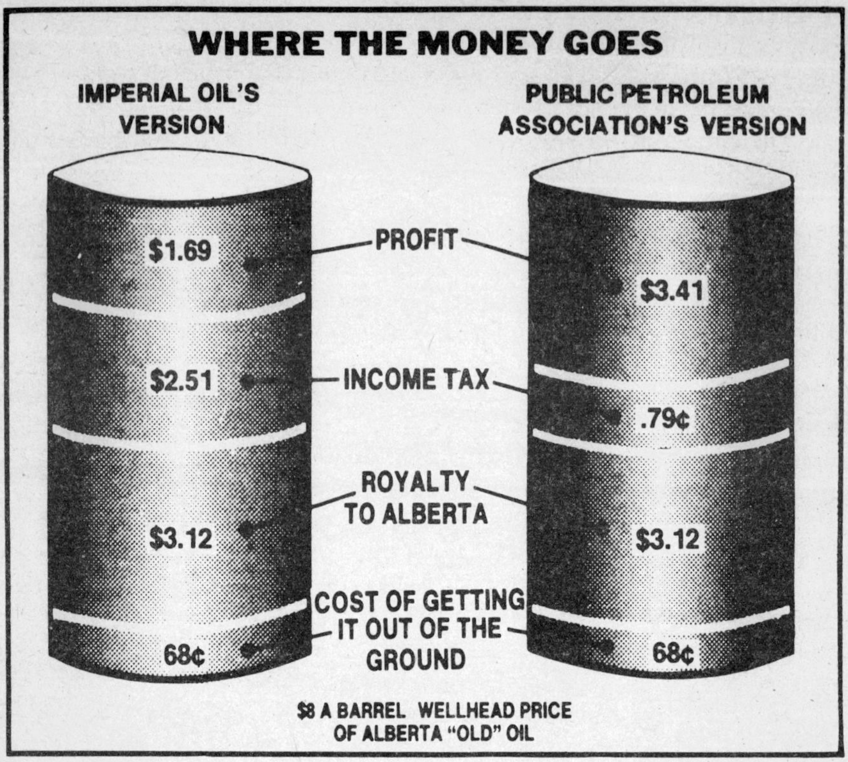
WHERE THE MONEY GOES
IMPERIAL OIL'S VERSION
PUBLIC PETROLEUM ASSOCIATION'S VERSION
$1.69
PROFIT
$3.41
$2.51
INCOME TAX
.79¢
ROYALTY TO ALBERTA
$3.12
$3.12
COST OF GETTING IT OUT OF THE GROUND
68¢
68¢
$8 A BARREL WELLHEAD PRICE OF ALBERTA "OLD" OIL

$3.41 a barrel.)[32]

Where has Imperial's much increased cash flow been going, if not into exploration for new oil? In 1975 the company spent $101 million on Syncrude,[33] a project that is developing a supply of petroleum in the high cost Alberta oil sands that the government has agreed will come on stream at the world price. Why has Imperial concentrated on Syncrude and not on exploration for new oil? In addition to the world price for the Syncrude product, Imperial has been motivated by tax incentives. Under the federal government's new earned depletion system, the bulk of Imperial's investments in Syncrude qualifies. Thus the company earned a depletion allowance at a rate of approximately $34 million a year on Syncrude in 1975, during the portion of the year in which the new depletion system was in operation. One firm of financial analysts estimates that in the second half of 1975, under the new system, Imperial was deducting forty cents per barrel on its conventional Alberta crude for tax purposes on the strength of its Syncrude costs.[34] Under the system the company is allowed to write off $4 for every $3 of its investments in Syncrude. In each of the next two years Imperial is expected to spend $200 million on Syncrude, two-thirds of Imperial's total exploration and development budget.

In addition to earning a guaranteed return on its Syncrude investment as a partner of three governments, Imperial's timing in putting the largest part of its current investment into Syncrude is significant. I believe that Imperial will continue to direct its investments in this way until Canadian domestic oil prices have reached the world level. Then Imperial can return to exploration for new oil. Imperial's present investment priorities are thus a considerable contributing factor in pressuring Canada to move to world prices.

There is considerable motivation for the petroleum industry to speculate on its present leaseholds and to delay drilling until the world price has been assured and until governments have been panicked into holding down royalties as an added incentive to the companies.

Congressional investigations in the United States have turned up considerable evidence that suggests that the oil and gas industry in that country has been deliberately understating proven reserves and holding up production and exploration to drive prices up to the world level.

The U.S. House of Representatives Subcommittee on Oversight and Investigations has found evidence that a number of major natural gas producers have either delayed development of certain reserves or failed to increase the deliverability in certain fields. In addition, the U.S. Federal Power Commission has found that producers are holding substantial amounts of proven gas reserves in a producible shut-in category.[35]

In January 1976, *The Washington Star* carried a story quoting a special assistant to a House of Representatives Commerce subcommittee as saying that if the American Gas Association (the industry's voice) has understated natural gas reserves by as much as a subcommittee investigation revealed had been done in one part of the U.S., the reserve underestimate would amount to 102.7 trillion cubic feet—five years supply for the United States.[36] A November 1975 Library of Congress study concluded: "Serious underreporting by the natural gas producers to the Federal Power Commission of natural gas reserves has existed and continues to exist."[37]

A Wall Street Journal story concluded that between 30 and 45 billion barrels of oil recoverable through existing conventional techniques is being left in the ground by the oil companies until prices rise to make it worth their while to produce this "old" oil.[38]

The U.S. Senate Subcommittee on Multi-national Corporations and U.S. Foreign Policy, chaired by Senator Frank Church, recently produced a summary of basics on the petroleum industry for the use of its own members. The summary stated:

> Like so many figures, proven reserves come from the industry and are not subject to independent or outside verification. However, a consensus exists among knowledgeable observers of the industry that figures on proven reserves are generally understated.[39]

Canada, however, is proceeding to alter the whole structure of oil and gas prices without any serious investigation of what happened to estimates between 1973 and 1976. Surely it would make sense to follow up on the leads turned up in the U.S. Congressional investigations. After all, Canada's petroleum industry is simply the northern extension of the U.S. petroleum industry. There is no reason to believe that it plays the game any differently here than across the border. Faced with the unanswered question about the lowering of our reserve estimates by 60 per cent, how can we simply proceed with the game plan the companies have mapped out for us?

In a brief to the Royal Commission on Corporate Concentration presented in May 1976 on behalf of his company, Imperial's Chief Executive Officer, Jack Armstrong, described his company as so regulated by government that it is "like living in a goldfish bowl."[40] Perhaps this accounts for the company's great eagerness in the past to contribute very large sums of money to political parties in Canada. It doesn't hurt to be friends of those who are in power.

The hearings of the committee on multinationals in the U.S. Senate (the Church committee) has documented, for the first time, the enormous degree to which the Liberal and Conservative parties have been financed by U.S. multinationals. Corporation after corporation has appeared before the committee and acknowledged making political contributions abroad. Usually the companies have been making contributions in two countries: Italy and Canada. The Italian contributions have led to a major political scandal in that country. But the companies have all told the Church committee that there's nothing controversial about their contributions in Canada. Jack Blum, assistant counsel to the Church committee, was quoted in *Maclean's,* as saying:

> Canada has cropped up most often as the place where political contributions are made by the big corporations. When we ask the companies about it, they always say 'Well, its legal up there.' If I were a Canadian, that answer would not satisfy me.[41]

Both Imperial and Gulf made it clear that their contributions went to political parties in Canada that supported the present economic system— that is parties that do not include nationalization of the petroleum industry in their platforms. Exxon was quoted in *The Wall Street Journal* as saying that this meant there had been no contributions to the NDP. And, Jerry McAfee, former chairman of Gulf Canada, now chairman of Gulf International, also made the point that the company's political largesse did not extend to the NDP.[42]

Let's look at the amounts involved. Exxon admitted that Imperial had contributed an average of $234,000 a year to political parties in Canada. Gulf contributed an average of $140,000 a year. For Shell, the figure was $100,000. *Maclean's* speculated that in the election year of 1974 the contribution level could have been twice as high. The eight oil companies admitting to contributions offered figures amounting to $546,000 a year to Canadian political parties in non-election years. If the Liberals spent $6.2 million on the 1974 election and the Conservatives spent $3.9 million, oil company contributions made up perhaps as much as $2 million of this or 20 per cent of the total.[43] And, of course, Imperial's contribution was by far the largest, nearly half the total. It's charming of Jack Armstrong to talk about how heavily his company is regulated when Imperial was, in all likelihood, the biggest single source of funds helping to elect the present parliament of Canada.

Obviously not wanting its political muscle to show, Imperial has announced that it will discontinue its contributions to political parties in Canada.[44] Gulf is more forthright. Its annual report for 1975 announced that contributions will continue as part of the company's commitment to "supporting the democratic process and the enterprise economy."[45]

Clearly, disclosure of political contributions is something the companies don't relish. The 1975 provincial election in Ontario revealed this. In February 1975 a provincial law went into effect requiring disclosure of the source of contributions in excess of $100.

In the period from August 11, 1975 to January 1976 Imperial made no contributions to either the provincial Conservative or Liberal parties. Gulf gave the Conservative central campaign fund $4000 and donated $500 to each of seven PC candidates. The company contributed $2000 to the Liberal central campaign and $500 each to four Liberal candidates.

Jack Armstrong satisfied himself personally with a $400 donation to Bette Stephenson's campaign (she was his local PC candidate) and a $125 contribution to Darwin Kealey, the PC candidate in Carleton East riding.[46]

Perhaps it is not unreasonable to suggest that the mammoth oil company contributions towards the election of the present members of the House of Commons has influenced the political climate so that tough questioning of the industry is not pursued by the government.

The federal government has revealed a marked capacity to go along with the projects with which Imperial Oil is closely associated. In the days when Imperial pressured Ottawa for greater exports, the Department of Energy, Mines and Resources made pilgrimages to Washington to seek greater access for Canadian crude to the American market. Today it is Imperial that has the most to gain or lose from the decision to build or not to build the Mackenzie Valley Pipeline. Imperial is the largest leaseholder in both the Mackenzie Delta and the Beaufort Sea. Petroleum analysts differ widely on the drilling prospects in the Beaufort Sea. One group of analysts estimated, on the basis of confidential information provided by Imperial, that the shallow waters of the Beaufort would yield as much as 50 trillion cubic feet of natural gas. Other analysts are more pessimistic. It is clear, though, that since 1972 some members of the federal cabinet have been committed to the Mackenzie Valley project. The fate of that project will heavily influence the value of Imperial's shares in the market. At the end of 1972 when confidence in the exploration in the Mackenzie Delta was high, Imperial shares were selling for $50.75. In the spring of 1976 they sold for $23.

The Liberal government and Imperial appear tied to each other in their support of the pipeline, whether or not the project makes sense for Canada. This may well explain the reluctance of the government to proceed first with Polar Gas, a project backed by greater proven gas reserves, that poses fewer environmental risks, would be less disruptive to native communities and would be an all Canadian system.

What is true of the Mackenzie Valley Pipeline is also, of course, true of

Syncrude, a project in which Imperial and the federal government are partners.

Imperial has been the clear pace setter in the petroleum industry in Canada. Its influence has been decisive in establishing the priorities for Canadian petroleum policy.

Establishing new, Canadian priorities for the petroleum industry that are geared to meeting the needs of this country's consumers and industry cannot be accomplished as long as Imperial takes its direction from an American head office. Bringing Imperial under public ownership as a Crown corporation is an essential first step in the establishment of a rational Canadian energy policy.

1 J.A. Armstrong, "Address to Annual Meeting of Imperial Oil Ltd. Shareholders," Toronto, April 26, 1976.

2 *The Globe and Mail*, July 9, 1970.

3 Imperial Oil Ltd., *Annual Report 1975*.

4 Imperial Oil Ltd., Supplement to *Annual Report 1975*.

5 *Imperial Oil Review*, 1976, No. 1.

6 George de Mille, *Oil in Canada West: The Early Years*, Calgary: Northwest Printing and Lithographing,1969, 145.

7 *Ibid.*, 146.

8 Touche, Thomson, and Yeoman Investment Consultants Ltd., *Research Report on Imperial Oil Ltd.*, Calgary, 1975.

9 *Ibid.*

10 *Ibid.*

11 *Ibid.*

12 *Ibid.*

13 *Ibid.*

14 *Maclean's*, April 5, 1976.

15 *Hockey Night in Canada,* CBC TV, February 7, 1976. This commercial message appeared on *Hockey Night in Canada* throughout the autumn-winter 1975-76.

16 Imperial Oil Ltd., *Annual Report 1960.*

17 Imperial Oil Ltd., *Annual Report 1963.*

18 Imperial Oil Ltd., *Annual Report 1969.*

19 *The Globe and Mail,* July 9, 1970.

20 Imperial Oil Ltd., *Annual Report 1971.*

21 Imperial Oil Ltd., *Annual Report 1972.*

22 Canada, Department of Energy, Mines, and Resources, *An Energy Policy for Canada—Phase I,* Ottawa: Department of Energy, Mines and Resources, 1973, Vol. 1, 104.

23 *Ibid.,* 101-2.

24 W.J Levy Consultants Corp., *Emerging North American Oil Balances: Considerations to a Tar Sands Development Policy,* New York: 1973.

25 For a discussion of the oil price change, see James Laxer, *Canada's Energy Crisis,* Toronto: James Lorimer and Co., 1975.

26 National Energy Board, *Report to the Honourable Minister of Energy, Mines and Resources in the Matter of the Exportation of Oil,* Ottawa: Information Canada, 1974.

27 Canada, Department of Energy, Mines and Resources, *An Energy Strategy for Canada,* Ottawa: Department of Energy, Mines and Resources, 1976, 80, 114.

28 *Ibid.,* 64.

29 *Ibid.,* 134.

30 W.J. Young, "Address to Winnipeg Society of Financial Analysts," Winnipeg, March 4, 1976.

31 Imperial Oil Ltd., *Annual Report 1975.*

32 This figure is derived from a calculation reached by applying current tax regulations to Imperial Oil's rate of production in 1975.

33 Imperial Oil Ltd., *Annual Report 1975.*

34 Baker, Weeks of Canada Ltd., "Imperial Oil Ltd.," Progress Report (PR-C, 5/7/75).

35 *Report by the Subcommittee on Oversight and Investigations of the Committee on Interstate and Foreign Commerce,* House of Representatives, 94th Congress, 1st Session, Nov., 1975.

36 *The Washington Star,* Jan. 21, 1976.

37 *Library of Congress Report,* Nov., 1975.

38 *Wall Street Journal,* Dec. 18, 1975.

39 Subcommittee on Multi-National Corporations of the Committee on Foreign Relations, U.S. Senate, *Multi-National Corporations and U.S. Foreign Policy,* Pt. 4, 198.

40 *The Toronto Star,* May 6, 1976.

41 *Maclean's,* Feb. 9, 1976.

42 *Ibid.*

43 *Ibid.*

44 *Ibid.*

45 Gulf Oil Limited, *Annual Report 1975.*

46 Political Contributions in excess of $100 to Provincial Party Offices, Candidates and Riding Associations, Statements filed with Ontario Government, Aug. 11, 1975-Jan. 18, 1976.

The Seventh Sister
DOCUMENTARY
The Banks and the Power Structure

INTERVIEW BY DAVE GODFREY
WITH MR. HOOLE, JULY, 1976

In late 1975, a man we shall call Mr. Hoole changed banks. An ordinary enough experience. And yet behind this everyday action lies quite an extraordinary story. The names have had to be changed for quite obvious reasons; the facts have all been attested to. Mr. Hoole changed banks because he was convinced that one of the major oil companies had conspired with bank officials against the independence of his firm, a heating-fuel distribution firm that had been in the family for three generations, a firm that has an annual sales volume of over three million dollars.

One thinks immediately of "integrity"—exactly as an old-fashioned, traditional concept—when meeting Mr. Hoole. His offices are well-lighted, but plain, with steel gray desks spread about in precise, unchanging order. The firm's name is one which the downtown inhabitants of the city have known for ages; it is seldom advertised, but one knows it from seeing the red and gray trucks pass by, from seeing the storage tanks alongside a wide old street. It is part of the city.

And Mr. Hoole himself seems a little surprised that he is upset. He is, after all, carrying on a tradition. People rely on him. Not because of an Esso-man image repeated and repeated on television, but because of direct contact. Hoole's helps provide heat for a good deal of the city; its procedures for dealing with inspections and repairs and night calls are well established. Reliable. Part of the order of the city.

And he is not happy that he has to claim anonymity for the purpose of this interview. You can see that it does not sit well with him. He is used to speaking the truth. He belongs to a good club. Though these things are not spoken of, he knows that he has a reputation for all the old-fashioned virtues: duty, truthfulness, service. He is a part of the community that can be relied upon.

In one sense the amounts of money under discussion are quite large. Measured in terms of jobs, and employees that have been with the firm for twenty or thirty years, and truck drivers, and furnace repairmen. In another sense, a multinational sense, what has caused the shock is miniscule. One working day's profit. A fraction of a per cent in terms of market-saturation. A president's salary, with stock-options still to come.

"And what happened when you went to the other bank?" I say. "Because that to me is the key. I don't mean that I do not believe you, but if a good lawyer were handling the other side, I am sure that what he would concentrate on is the timing of the application, the possibility that this was all merely an error, a flaw in management skill, not a form of pressure of any kind."

And I watch his reaction closely, because the banks, the large, core-of-the nation, Canadian banks do not do what Mr. Hoole is suggesting they do. Of course, they loan money more easily to American multinationals than to Canadian businessmen, of course they even encourage such purchases, and from time to time even find possibilities, or pass on hints, information, and suggestions to their multinational customers. And of course, in a sum-of-individual-interests world there is a case to be made for such activity. And Mr. Hoole would probably be fully in favour of the philosophy underlying it. But something has upset him. He is aware that the business world is not a world of Boy Scouts. But there are limits. And yet, even I have my doubts, at a social instinct level, that this bank has done what he says it has. Because, summing it up, their action represents a stage of economic domination from abroad which is simply not tolerable. By any level of the community.

The Hooles have dealt with banks for three generations. They have watched presidents come and go and names hyphenate and unhyphenate. They service a number of bank branches with heating fuel. But they have not been great borrowers. Integrity is measured for them not in the size of their offices or the extent of their advertising campaign or the capacity of the company jet, but in the smallness of their debt to capital ratio. In fact, for many years, they have actually loaned money to the bank. In the winter, as their accounts came due, they would deposit funds with the bank and build these up until the following late summer when they would once again replenish their storage tanks, reduce to a low cash balance, and build up their accounts receiveable.

And they received offers to sell out. The city is a flourishing one and the customers long-established. The employees stay on and on and the profit is reasonable. Hoole's is large enough that it can buy its supplies from several of the big seven oil companies in any given season and the laws of the market place still seem to apply. When the big seven come around, or one of their Canadian employees talk to Mr. Hoole in the club, they are very polite and circumspect about making their offers, or suggesting that such offers are a distinct possibility. They talk his language. Or at least try to. It is not Imperial that has been most persistent. It is, let us say, Big 7 Oil.

And Mr. Hoole knew that the President of Big 7 Oil sat on the Board of Directors of his bank. But there are rules to this game. The gossip of friends and acquaintances is one thing; the passing on of confidential information to a competitor quite another.

The story begins in 1972. From 1955 until 1972 Hoole's had a line of credit of $500,000. But it wasn't used. In 1972 they decided on a change of policy, investing funds on hand and borrowing for the fall cash requirements. They went to the bank to talk over the loan and present financial statements. Present were Mr. G. Sutherland, the manager, and Mr. Black, the loan officer. The manager surprised Mr. Hoole by asking if Hoole's were interested in buying any other firm and indicating that the bank had a department to handle such matters. Mr. Hoole told him that if he were ever interested he would be in touch, but that he had come to arrange a loan.

A few days later, in his office, he received a telephone call from Mr. Black, who stated that they had been in touch with a firm very interested in buying Hoole's.

"I told him that this was quite a reversal; that when I was there they had asked me about buying not selling. I told him very clearly that we were not interested. But of course the line of credit was approved, accounts receivable assigned, and the usual guarantees and documents signed."

In 1974, shortly after Labour Day, Mr. Hoole arranged a meeting with bank officials to increase the line of credit to $1,200,000 because of the large increase in oil prices. Financial data was supplied and Mr. Hoole was assured that there would be no problem. There was a new credit officer now, a Mr. Dunn, reporting to G. Sutherland.

But weeks went by and nothing was heard from the bank. Two large suppliers were pressing Hoole's for payment, but nothing could be done.

Finally, a lady from the bank called Mr. Lash, the Secretary of Hoole's, and told him that the loan had been approved. Cheques for over $700,000 were issued. Both of these cheques were cashed by October 25th, with the approval of the bank and without informing Hoole's of any problems or difficulties, bringing the draw on the line of credit to $1,200,000, the limit requested.

At this point the drama began. Mr. Dunn telephoned Mr. Hoole, on the afternoon of the 28th, and only then informed him that there was an "overdraft" of $200,000 which must be cleared up before the line of credit could be proceeded with. Mr. Hoole was shocked.

At the time, of course, he could only act. With limited experience of lines of credit, he was frightened by the ultimatum. Visions of bankruptcy flashed through his mind, of creditors petitioning the firm into receivership after all these years. But putting the pieces together later on caused a deep anger within him. "I know now, of course, that an honest banker, if there really had been an overdraft, would have called me in and talked the matter over, maybe charged a little more interest, as they helped us work it down, but they would never have produced such an ultimatum.

"The problem is that there's only one logical interpretation that can be put on the bank's action. Big 7 Oil had a member on the Board. That member knew G. Sutherland. Sutherland and Dunn met together to stall the application. Then, when the cheques were out and Hoole's supposedly in trouble, they called us in. The next step in the scenario no doubt was a friendly suggestion that Big 7 Oil would step in to provide some cash, in the form of equity purchase, to tide Hoole's over this cash crunch."

Mr. Hoole becomes angry when he tells this part of the story, a justified anger.

"It's obvious. These bank officials had used a trusted relationship, used their confidential access to our financial data, to provide a large, non-Canadian firm with an opportunity to buy us out. I was shocked."

Luckily Mr. Hoole was a little more than shocked. The firm had never before been in an overdraft position and he did not want it stated now that it was. Rathei than fight the case through the upper echelons of the bank, he turned to private family funds. Within a few days the "overdraft" was covered. For the first time a slight chuckle arises deep in Mr. Hoole's sense of propriety and lightens the office.

"That's how we settled the so-called overdraft," he says. "The trust company had over a hundred and fifty thousand dollars on Mr. Dunn's desk within twenty-four hours. I would have liked to see Dunn's face when he reported that to Sutherland."

The bank's officers underestimated Mr. Hoole in two ways. The figures they were working with were only figures for the operating firm. The holding company of the Hoole's was as solid as Mount Royal. They also underestimated his integrity. Rather than being bluffed, he called their game completely. Rather than being intimidated, he went directly to the General Manager of the bank and revealed the whole story. But Mr. Hoole could receive no satisfaction from the bank, no admission that fraud and misrepresentation had taken place on the part of the bank. So he withdrew the account. And transferred it to another bank.

"To answer your question," he says, with a smile, having seen my kind of doubt before, "the other bank gave us a greater line of credit, and lowered the interest rate—without even asking me. That's one final proof. And the other, of course, is the way Dunn and Sutherland lied when the case was examined. If it was just a flaw of management, a 'breakdown in communication' as the bank termed it, they would have had nothing to hide. Now, as far as I'm concerned, the whole bank has something to hide."

A happy ending, of sorts. The family firm remains a family firm and hundreds of customers know nothing about the drama in high places. But how many firms have the resources, financial and of integrity, that Mr. Hoole could draw upon? How many takeovers were consumated at this point? when the friendly firm arrived to cover the overdraft and relieve the situation without embarrassment to anyone. How many?

L
MG

2

FUELISHNESS IN OTTAWA: THE FAILURE OF CANADIAN ENERGY POLICY

MEL HURTIG

Edmonton publisher Mel Hurtig has been a leading critic of the Canadian-American relationship and of government policies on natural resources. His writings appear regularly in newspapers and magazines across Canada. Hurtig Publishers have produced many titles which have won the Governor General's Award and other prizes. Mel Hurtig was named Canadian Book Publisher of the year for 1974. He is a founding member of the Committee for an Independent Canada and a former National Chairman of that organization. During the last few years he has served in an advisory capacity with many Canadian government and public interest bodies. He is on the National Executive of the Public Petroleum Association.

3

Mel Hurtig

The Mismanagement of Canada's Petroleum Resources

For many years we Canadians have been the recipients of voluminous advice about our non-renewable oil and gas resources, advice that now has left Canada in very serious economic trouble. Most of the advice came from the foreign-owned major petroleum companies and from provincial and federal politicians who were close to them. The message was bold and simple: Canada is rich in energy resources so let's export as much as possible to American markets as quickly as possible. The advice, year after year, was consistent and persistent, right up to the moment when international oil prices quadrupled in the fall of 1973.

Any proper evaluation of the management of Canada's energy resources must begin with an account of the incredible advice Canadians received from those who helped formulate the critical decisions about Canadian energy policy. Here is a short but representative summary:

*In 1969, the National Energy Board forecast that by 1985 Canada would be exporting over 900 million barrels of oil every year to the United States, gradually increasing to one and a half billion barrels in *annual* exports by 1990.

*One year later, the Canadian Bankers' Association forecast that, by 1985, Canada would be exporting *at least* 750 million barrels of oil annually to the U.S.

*In mid 1971, the then Minister of Energy, Mines and Resources, the Honourable Joe Greene, gave his memorable speech to the Canadian Institute of Mining and Metalurgy in which he stated that Canada had 932 years of natural gas , and 390 years of oil left in the ground.

*The next year, 1972, the Minister of Industry, Trade and Commerce, the Honourable Jean-Luc Pepin, agreed with statements made by Ronald Ritchie, then vice-president of Imperial Oil and more recently Robert

Speaking of Petroleum

It would be crazy to sit on it. In maybe 25 to 50 years we'll be heating ourselves from the rays of the sun and then we'd kick ourselves in the pants for not capitalizing on what we had when gas and oil was a current commodity.

JEAN LUC PEPIN, *August 10, 1970.*

Stanfield's man in Ottawa, that Canada should be selling its reserves of oil and gas as soon as possible. This is what he said:

> In maybe 25 to 50 years we'll be heating ourselves from the rays of the sun, and then we'll kick ourselves in the pants for not capitalizing on what we had when gas and oil was a current commodity.

*Early in 1973, former energy minister Donald Macdonald described Canadian petroleum reserves as "more than adequate."

*In 1970, because of our supposed huge reserves, and since Canada was so adequately protected in domestic requirements for *decades* into the future, the National Energy Board and the Trudeau cabinet agreed to increase exports of natural gas to the U.S. by some 50 per cent, an additional 6.3 trillion cubic feet.

*In every year from 1965 through 1973, because of our huge mythological reserves, and because growing domestic demands were supposedly so well protected, Canada increased its exports of crude oil to the United States. Annual exports more than quadrupled from 1965 to 1973.

*In the spring of 1973 Alberta Premier Peter Lougheed had this to say:

> Alberta cannot agree to having to keep its oil in storage for tomorrow in order to serve a slow-growing eastern Canadian market, or to have its oil available for some kind of emergency.

*In the fall of 1973, Mr. Lougheed talked about the Alberta tar sands:

> Because of the enormous size of the Alberta oil sands, obviously the market has to be essentially an export market, in the United States, possibly Japan.

*And two months later:

> Rapid development of the oil sands will allow Canada to be in a strong position to be able to supply additional exports to American markets.

*Just in case someone didn't get the message, here is the Alberta premier on December 4, 1973:

> The magnitude of the Alberta oil sands is so great it can supply oil surplus to Canadian needs for export to the United States. Alberta would like a Canadian-United States treaty by which . . . Alberta's oil would be sold in the U.S.

*In 1973 our friends at Shell Oil had this to say:

> There is ample Canadian crude to meet Canadian requirements and also to continue to provide substantial exports. . . .

There are scores of similar examples. Now, as a consequence of all this enthusiastic advice and optimistic predictions from oil company presidents, from our federal and provincial politicians and from officials of the NEB and Alberta's Energy Resources Conservation Board, where do we stand today?

In 1976, for the seventh year in a row, Canada will experience a drop in proven oil reseves. Not one single reliable forecast predicts other than a continuing severe downward trend in these reserves.

Now, only a few short years after Mr. Greene told Canadians that we had a 390 year supply of oil, Canada has about nine years of proven conventional reserve. With very good luck, as much as two billion barrels of new oil in the western sedimentary basin may yet be discovered. Two billion barrels would meet Canadian domestic requirements for only two to three years in a few years time. The Pembina field in Alberta was the biggest and most productive oil field in Canadian history. To maintain our present reserve levels we would require the discovery of the equivalent of a new Pembina field *every year*, year after year. The last major oil strike in Alberta was nine years ago!

By 1980 Canadian domestic demand will be about 2.3 million barrels per day. Even with the discovery of an additional billion barrels of oil in western Canada, in the Arctic, or offshore, we would have the equivalent of only one year's additional supply.

In 1973 we increased our oil exports to the U.S. to 1.3 million barrels per day once again because future domestic requirements were supposedly so well protected. Only two years later, in 1975, Canada became a net *importer* of oil (that is, Canada began importing more oil than it was exporting)—expensive, *very* expensive Middle East and Venezuelan oil, expensive and insecure.

In 1973 crude oil imports cost Canada $941 million. In 1974 this cost tripled to $2,645 million. In 1975 the cost was over $4 billion. These import costs have to be weighed against the fact that now Canadians are reducing oil exports to the U.S. because, late in 1974, even the NEB and Donald Macdonald and Peter Lougheed began to realize what some of us had been saying for six or seven years now—that Canada has been very quickly running out of oil.

As Canada's status as a net oil importer inevitably increases in the future, we will suffer increasingly large, multi-billion dollar oil trade deficits. By 1985, every day, Canada will be importing one million barrels of offshore oil to service the Canadian market *west* of the Ottawa Valley, in addition to our normal eastern Canadian requirements. Our oil deficit during the next decade will cost us $35 billion or more.

Almost the same thing is happening in respect to natural gas. While Canada continues to export one trillion cubic feet of accessible, high quality natural gas to the United States each year, our own country will soon be running short. Joe Greene's famous 923 year supply of a few years ago now stands at about a fifteen to twenty year supply today. Parts of Canada have already experienced natural gas shortages—shortages which will certainly worsen year after year. And even where supply is available it will cost *at least* six times as much as Canada charged for most of the gas exported to the U.S. in recent years.

A further component of the bleak supply picture is that there is no way the tar sands can be developed quickly enough to alter Canada's sad energy picture and even with the discovery of an elephant-size Arctic reserve of the magnitude of Prudhoe Bay, we would still be faced with a *long-term* inability to meet our domestic oil requirements.

How does this record of false oil company and government predictions and the resulting shortages fit into the overall state of the Canadian economy?

In 1974 Canadian imports totalled $31.5 billion, an all time record. In the same year, the country's current accounts deficit of $1,877 million was another all time record. In 1975 Canada's current accounts deficit reached a colossal $5.4 billion. 1976 will show another staggering deficit.

In short, Canada is in trouble, big trouble. It is trouble that will affect everyone in the country in a profound way for many years into the future. The mismanagement of Canada's reserves of non-renewable energy resources has been colossal and the long-term effects on the Canadian standard of living will be equally gigantic.

In assessing how this has happened, we must look at the remarkable performance of the foreign major oil companies, both individually and collectively, and their so-called "Canadian" Petroleum Association.

The CPA has operated, at both the provincial and federal levels of government, one of the most powerful and well-financed lobbies in Canadian history. At the same time, the CPA has maintained close ties with the U.S. State Department and the U.S. Department of Commerce.

For almost all of the past quarter-century, most of Canada's federal and provincial petroleum policies have been evolved in direct response to policymaking input from large foreign firms, most of which were in turn part of the international oil cartel which for decades controlled almost exclusively the supply and pricing of the non-Communist world's petroleum production.

In Canada these foreign "majors" own 91 per cent of the petroleum industry, make 95 per cent of the profits, and control 99.9 per cent of total petroleum refining.

Three decades of taxation and incentive schemes, production rates,

"Put A Tiger in Your Tank"

Imperial Oil Review

OCTOBER 1965

the degree of domestic refining and upgrading, and Canada's ill-conceived "national oil policy" until recently were all formulated mainly on the basis of evidence presented by large foreign oil companies to the NEB, the Alberta ERCB, and provincial and federal cabinets.

These firms spend tens of millions of tax-deductible dollars advertising their good coporate citizenship and explaining how important they are to Canada. They take out full-page newspaper advertisements and sponsor prime-time television programmes. They sponsor television hockey games, give away medals depicting the prime ministers, employ the best public relations staffs, spend millions supporting Canadian political parties and individual politicians, and in general have a pervasive influence on Canadian society.

Without exception, this small handful of multinational subsidiaries—this small group of eight enormous foreign companies that the Carter Commission said received 85 per cent of the Canadian petroleum industry's tax privileges—are, *without exception*, firmly controlled outside of Canada by their parent companies.

In Canada these foreign-owned and controlled subsidiaries make special efforts to present themselves as solid citizens. Imperial Oil (70 per cent owned by Exxon) has quite successfully misled a large percentage of Canadians into believing that it is a Canadian company. In actuality both the long term and even much of the day-to-day decision-making is tightly controlled from outside Canada.

At this point a further and seldom considered factor must be understood. The U.S. petroleum industry has long been regarded by the U.S. State Department, and the White House, as being an instrument of American foreign policy. In one instance the Eisenhower administration even dropped anti-trust action against the cartel so as to avoid embarrassing the industry abroad, thus endangering future U.S. profits and supplies. Many similar examples of the U.S. government-petroleum industry relationship surfaced during the 1974 Congressional hearings held as a result of the "energy crisis". The long history of manipulation of the U.S. government by the petroleum lobby has been well documented.

Difficult as it may be to accept, the truth is that, for most of the period since the Leduc discovery in 1947 until very recently, Canadian federal and provincial petroleum policies have been to a great extent the result of similar manipulation.

No examination of the history of oil and gas policy in Canada can arrive at any other conclusion but that decisions were designed so as to have one foremost objective: *to maximize exports to, and profits in, the United States.*

Anyone taking the time to go back and study the input to the NEB and Alberta's ERCB will have no difficulty gaining a discouragingly accurate perspective. Time after time, in well-prepared briefs presented by high-priced, capable and well-equipped attorneys, geologists and engineers, the industry seriously misled the regulatory bodies. And over a period of years the industry virtually monopolized the hearings. In this long process the relationship betwen the regulatory bodies, which were supposed to protect the public interest and keep their governments informed, and the industry was such that the record clearly shows a remarkable failure by Canadian civil servants to exercise their responsibilities to their country.

The worst record in decision-making is that concerning reserves as discussed above. But other decisions have resulted in absurd policies also affecting the economy as a whole.

One of these areas has been taxation. For years now the *effective* rate of federal income tax paid by the oil companies, some of the wealthiest and most profitable companies in the world, was between 5 per cent and 10 per cent of declared profits. For some it was even less. And despite well-published claims to the contrary, the petroleum industry has been a more than excellent profit earner. For *the past few years* the industry has been making much larger profits than it ever dreamed possible in 1972 even in its most extravagant dreams!

In Venezuela and in the Middle East today, the same petroleum companies receive between $.75 and $1.25, net, on a barrel of oil. In Canada, on $8.00 oil—remember, the price is going to $9.50 by January, 1977—they are even now receiving between $2.75 per barrel and $3.00 per barrel. And they are hollering all the way to the bank.

Not counting refining profits, distribution profits, pipeline profits, and petrochemical profits, the industry's 1975 profits before taxation, after all expenses including royalties, were two and half billion dollars—the highest in history.

Contrary to public oil company statements, even if only 10 per cent (a *very* low figure) of the value of presently proven reserves eventually accrues to the petroleum companies, the ratio of earnings to capital invested, or even to capital employed, will be very high.

The failure to encourage the early development of a Canadian petrochemical industry is another result of the influence of foreign oil companies on the decision-making process. The foreign-dominated industry could hardly be expected to develop a healthy Canadian petrochemical industry when Canadian oil and natural gas could be readily obtained for the parent U.S. petrochemical plants. The resulting loss to Canada in unrealized petrochemical revenue, job creation, tax revenues and spin-offs can be measured in further billions of dollars. Similarly, the resulting obligatory imports of petrochemical products from the U.S. have been damaging to the Canadian balance of payments.

Another failure has been the neglect of domestic energy research and development based on Canadian priorities. With tens or even hundreds of billions of barrels of potentially recoverable tar sands oil, and with the important initial research work already undertaken by the Alberta Research Council, Canada could have been the world leader in *and* owners of the tar sands separation and in-situ technology. Had such research proceeded, using capital injection from the petroleum industry and/or government, the tar sands could have been developed in an unhurried, environmentally protective manner allowing maximum revenue to remain in Canada. It hardly need be said that such a game plan was not encouraged by the foreign petroleum industry which had less secure and highly profitable offshore oil to sell.

The all-inclusive costs of hurried Arctic exploration, field development and pipelines, of rushed tar sands plants, and of the $12 billion James Bay project, and the environmental consequences of their fast planning and construction, are also a direct result of miscalculations in the past. For a nation dangerously close to the point of no return in the degree of foreign ownership and control we have already accepted, these new resource developments move us still closer to the point of surrendering even the opportunity to reverse the thirty year process during which we became the world's foremost example of a "developed" economic colony.

In talking about the massive foreign-controlled energy projects which Canada so readily accepts, it must be understood that the greatest mythology is that these projects require huge amounts of foreign direct investment.

At the end of 1973, the total foreign, long term investment in the entire petroleum and natural gas industry in Canada was about $9.5 billion. To be generous, it can be estimated that by the end of 1976, foreign petroleum companies will have brought into Canada some $12 billion. On western Canadian production alone, the revenue of the petroleum industry, at the end of 1976, will exceed expenditure by *at least* $7 billion. This means that by the end of 1976, on a total investment in western Canada of $31 billion, the foreign companies will have imported roughly 36% of the capital, while making 95% of the sales and 95% of the industry profits. The after-tax profits on production alone will have yielded a very minimum of $4 billion in return; or a return on foreign-invested capital somewhere between the rates of 30-35%. I remind you that this does not count foreign-controlled profits in refining, distribution, petrochemicals and pipelines.

How much capital will be required to develop the petroleum industry

New York Oil Consultant Walter J. Levy: Advisor to Ottawa and Edmonton

Although very few Canadians have ever heard of him, Walter J. Levy has influenced Canadian petroleum policy to an extent surpassed by few men.

An American citizen, a former senior officer of the Office of Strategic Services (the predecessor of the CIA), a former employee of the Office of Intelligence Research in Washington, a long-time presidential and state department special consultant, Levy is a New York oil expert who has been an intimate adviser to both the federal and Alberta governments during the past 15 years.

Levy has been instrumental in convincing both Alberta and Ottawa that Canada had plenty of oil and that if we behaved properly, the US would do us a favour by taking some of the surplus off our hands.

It was Levy who helped convince Canadians not to build a pipeline to Montreal to service their own eastern markets.

It was Levy, in his New York office, who actually drafted the 1961 Canadian National Oil Policy that helped American petroleum companies export Canadian crude to their US refineries while making the highest possible profits importing offshore oil into eastern Canada.

And it was Levy who advised the Alberta government in 1973 to proceed with rapid development of the Alberta oil sands for export to the United States. As a special, part-time consultant to Peter Lougheed, he received $60,000 a year.

Levy's consistent advice has been that Canadians should entrust the development of petroleum resources to US based petroleum companies. Yet in advice to his own country on oil policy, on the question of Arab petro-dollars entering the United States, Levy stated:

"I very much doubt whether we or the citizens of any other developed country in the world would allow non-residents to come in and buy up control of our vital industries and natural resources."

It is perhaps not surprising that this man was awarded a special plaque by US Secretary of State Dean Rusk for his distinguished service to the United States and that he received the President's certificate of merit.

in Canada in the future? Recently, the Independent Petroleum Association of Canada suggested that between $50-70 billion will be needed in the next ten years. A.H. Ross, who was speaking on behalf of the Independent Petroleum Association, said that only two-thirds of that $50-70 billion can be raised internally in Canada. But, during the past decade, the entire expenditures of the petroleum industry, including geophysical expenditures, drilling, land acquisitions and rentals, field equipment, secondary recovery and pressure maintenance, taxes, royalties, etc., was about $17 billion. If the projected industry figures are correct, then Canada could finance all petroleum development here without permitting the kind of massive foreign ownership we have been so hospitable to in the past, even if the industry were to double and perhaps even triple its activities of the past decade.

The consequences of Canada continuing its foolish economic branch plant/resource colony pattern of the past will certainly be extremely high unemployment, a back-breaking perpetual national debt, and a nation increasingly unable to control effectively its own monetary, fiscal and social policies.

It is, of course, wrong to lay the entire blame for past mistakes at the door of internal or external foreign influence on Canada's decisions without recognizing that the real responsibility rests in one place and one place only, with our federal and provincial political leaders.

For years we've had no national energy policy. Today after all the debate, there is still no national energy policy that properly safeguards the national interest, and there is no prospect for the evolution of one. The reason, never admitted by Mr. Trudeau and never hinted at by our energy ministers, is quite simple.

It is in fact quite impossible to have a national energy policy without a national industrial policy. Since the end of the Second World War Canada has not had a national industrial policy, and Alastair Gillespie, when minister of Industry, Trade and Commerce, told us clearly there were no plans to formulate such a policy.

Again, the reason is really quite simple but never publicly admitted in Ottawa. Because most of the vital economic decisions relating to the opening and closing of plants, the levels of wages and dividends, the direction of exports and the sources of purchases, the levels of research and development, and the degree of processing and refining are made largely *outside* of the country by the residents of other countries, the ability to create a national industrial policy based on Canadian priorities has been beyond our political leaders and civil servants. The current and increasing level of foreign ownership in Canada makes both an effective national industrial strategy and sensible national energy policy impossible to formulate.

Canadians have been badly misled and the true consequences are only now beginning to be understood. After ten years of studying the activities of the petroleum industry in this country I can only conclude that foreign companies and their senior employees have done great damage to our country, damage that can be measured in terms of scarcities, real increasing deprivations and at a cost of many tens of billions of dollars.

Canadian business executives who mislead regulatory authorities and governments into making vital decisions harmful to Canada should be prosecuted and, if found guilty, should be heavily fined and jailed.

Foreign companies guilty of the same offence should be nationalized without hesitation. Their executives should be heavily fined, jailed and then expelled from the country.

My last point has to do with our own political leadership. The records of Joe Greene, Donald Macdonald and Alastair Gillespie, our recent energy minister, are dismaying. Pierre Trudeau and Peter Lougheed have both been badly served by their own energy advisers and by the industry. Until very recently the National Energy Board has had a disgraceful record. Even today the Alberta Energy Resources Conservation Board performs almost as if its sole function was to act as a spokesman for the aspirations of the foreign-dominated industry.

While I certainly do not believe that it is already too late for Canada to regain control over its industry and natural resources, I am also firmly convinced that neither the Trudeau government in Ottawa nor the Lougheed government in Alberta will make any important departures in the future from the long-term historical trend of the ever-increasing foreign ownership of Canadian industry and natural resources.

You and your children and their children will all pay the enormous costs in the future. And the economic costs, heavy as they will be, will be the least important.

The Syncrude Team

(l-r) R.B. Galbreath, Canada Cities-Service Ltd.; Jerry McAfee, Gulf Oil Canada Ltd.; J.A. Armstrong, Imperial Oil Ltd.; Frank Spragins, Syncrude President

FRANCOIS BREGHA

Francois Bregha has been actively involved in energy issues for the last four years through his work with various public interest groups. A graduate in Political Economy from the University of Toronto and in Environmental Studies from York University, Francois Bregha has focussed most of his efforts on Canadian natural gas policy. He is currently working for the Canadian Wildlife Federation on presentations to the National Energy Board Hearings on the proposed Mackenzie Valley Pipeline.

4

Francois Bregha

Canada's Natural Gas Industry

1 STRUCTURE

In contrast to the oil industry which has achieved a high degree of vertical integration, the Canadian natural gas industry has remained divided into four more-or-less discreet sectors; production, gathering, transmission and distribution. With the exception of the producing sector, there is little overlap between the oil and gas industries. However the oil industry's strategic control over gas production gives it a very considerable influence over gas policy. Oil firms account for the bulk of Canadian gas production because the gas discoveries have historically been the by-product of oil exploration. This comes about because it is often impossible to foretell which hydrocarbon a particular geological structure may yield, a situation which remains to this day, particularly in the frontier regions. The oil producers have traditionally shied away from a more substantial participation in the gas industry because it offered a lower profit potential and because of the presence of government regulation in the transportation and marketing of natural gas, a presence attributable to the inherently monopolistic character of this fuel.[1]

If the Canadian natural gas industry has not become vertically integrated, it has nevertheless established numerous linkages among its various sectors. These linkages historically have tended to militate against the free interplay of market forces.

One example of these linkages is provided by Norcen Energy Resources Ltd. (formerly the Northern and Central Gas Corporation Ltd.), one of the three Ontario distribution utilities. Norcen holds interests in gas exploration and production through Canadian Industrial Gas and Oil Ltd.(CIGOL), in transportation through TransPrairie Pipelines Ltd., and the Champion PipeLine Corp. Ltd., and in distribution through Columbia Natural Gas., the Greater Winnipeg Co., Gaz Metropolitain Inc., Beaver River Utilities Ltd., and it own extensive franchise areas in Ontario.[2]

For a crucial period during the fifties, three of the four largest western gas producers (Gulf Oil Canada Ltd., Hudson's Bay Oil and Gas Co. Ltd. and Canadian Delhi Oil Ltd.) held an important interest in TransCanada PipeLines Ltd. (TCPL). [3] Today, TCPL is controlled by Canadian Pacific Investments which holds 15.91 per cent of its shares.[4] Canadian Pacific's controlling interest in PanCanadian Petroleum Ltd. (87.2 per cent), one of the ten largest gas producers in Canada,[5] links TCPL to the production sector. TCPL is further tied to the other sectors of the gas industry by its connection with Home Oil Co. Ltd., which owns 1,186,000 of TCPL's common shares.[6] Home Oil produces both gas and oil, and is controlled by Consumer's Gas Corp. of Toronto, a distribution utility.

The most direct corporate linkages, however, are to be found in the export companies. The Alberta and Southern Gas Company Ltd. (A&S), the largest exporter of Canadian natural gas, is a wholly-owned subsidiary of the Pacific Gas and Electric Company of California as is its sister pipeline company, Alberta Natural Gas, which delivers gas from Coleman, Alberta to Kingsgate, B.C. to Pacific Gas Transmission (also a subsidiary of PG & E); the Canadian-Montana Pipe Line Company, for its part, is a wholly-owned subsidiary of the Montana Power Corporation. This affiliation has carried with it important consequences for the pricing of Canadian gas exports.[7]

Another linkage among the various sectors of the gas industry is provided by the establishment of consortia to build arctic pipelines (e.g. Canadian Arctic Gas Pipeline Ltd., the Polar Gas Project). Arctic Gas, or CAGPL, at one time had twenty-seven members representing the most important companies from every sector of the industry in Canada and the United States.[8] Such an alliance is not only an unparalleled concentration of economic power but serves as well to transcend the normal differences among the industry sectors and those between Canadian and American companies by giving all a common goal. Consortia such as CAGPL thus help to minimize cleavages among the domestic sectors of the Canadian gas industry while, at the same time, they create yet another transnational link between Canadian and American oil and gas interests.

It should be noted, of course, that interlocking directorships[9] and the mutual interdependence of the segments of the industry encourage a shared ideology even when no direct corporate affiliation exists. Thus the interests of the various sectors will sometimes overlap and lead to the joint advocacy of certain policies. In its brief to the recent National

Energy Board hearings on natural gas supply and demand, for example, Consumer's Gas urged that greater incentives be given to the producer, a recommendation repeated in the TCPL final argument to the Board.

The apparent pluralism of the gas industry is thus revealed as somewhat deceptive because it masks the numerous linkages among its sectors which work to reduce the natural internal tensions one would logically expect within the industry. These linkages promote the development of shared values and, hence, a certain unity of purpose which cannot be explained exclusively in terms of a common concern in encouraging rising gas consumption.

It is evident, of course, that in times of crises, such as the present one in which shortages are predicted for the near future, the "recognition of mutual interests" in increasing gas supply will constitute a powerful force in overcoming partisan considerations and will therefore strengthen even further the present bonds which already unite the various sectors of the industry.[10] It is interesting to speculate whether the threat of domestic shortages in Canada today may have been artificially induced by the producers in the expectation that such a strategy would unite the gas industry behind them in pressing for the early construction of a Mackenzie pipeline.

The Canadian natural gas industry is characterized by a concentration of economic power which the preceeding discussion has not fully revealed. TransCanada, for example, necessarily exerts an impact on policy by being the largest purchaser of natural gas in Canada and the only transmission line east of Alberta. Similarly, by virtue of its monopoly on gas transportation in the province, Alberta Gas Trunk Line Ltd. (AGTL) plays a significant policy role, a role which is now increasing through AGTL's diversification in the petrochemical industry and its participation in the 'Maple Leaf' project to bring Mackenzie Delta gas to market.

Nowhere is this concentration of economic power more formidable, however, than in the production sector of the industry, where the large producers' influence over policy is the direct result of their affiliation with the multinational world oil industry. The oil companies' penetration of the gas industry at the production level has given the gas producers a certain independence *vis-a-vis* the other sectors of the industry but more importantly, it has given the producers immense weight in the formulation of Canadian gas policy.

Although the term 'gas producer' is used extensively in any discussion of the Canadian gas industry, it constitutes a conceptual abstraction which obscures the fact that there is no independent gas producer among

Exxon's Foreign Earnings: Investment Pool for US

M.A. WRIGHT *(Chairman and Chief Executive of Exxon Co., USA)*: Exxon's foreign operations have in no way been detrimental to our ability to make investments in the United States. Our foreign earnings, plus depreciation and off-shore debt, have been more than sufficient to fully finance new foreign investments for many years. . . .

THE CHAIRMAN: On balance is your company and the other oil companies bringing dollars back to the United States or taking money from the United States?

MR. WRIGHT: Well, you know, over the years we have always prided ourselves as having a plus balance of payments to the United States. Now I recall in the middle sixties we were always proud to let it be known that we were contributing something like a half billion, $500 million a year toward a $2 or $3 billion (surplus) in the United States. One of our companies has been one of the important contributors to our favorable balance of payments because of bringing the profits and the cash flow from abroad back into this country.

THE CHAIRMAN: Would that be true for the past 6 months?

MR. MONROE *(Comptroller, Exxon Corp.)*: Would you accept an answer for all of 1973? It looks like we brought back $800 to $900 million from our operations in 1973.

THE CHAIRMAN: You brought that back?

MR. MONROE: We brought that much home after we met our investment requirement during the year 1973.

MR. WRIGHT: Of course, what is obvious, bringing this amount of money into the United States is part of the financing that is available to us operating in the United States.

SOURCE: U.S. Senate, Committee on Finance, Hearings, *Profitability of Domestic Energy Company Operations*, Washington: U.S. Government Printing Office, Feb. 13-14, 1974, 175,180.

EXXON CORP., CONSOLIDATED RESULTS—FOREIGN
(In millions of dollars)

	1964	1965	1966	1967	1968	1969	1970	1971	1972	Estimated 1973
Expenditures less depreciation:										
Foreign capital spending	562	545	645	931	918	1,039	1,084	1,141	1,267	1,437
Less foreign capital recovery	(345)	(363)	(377)	(424)	(445)	(467)	(464)	(607)	(597)	(660)
Net change in plant	217	182	268	507	473	572	620	534	670	777
Other foreign expenditures, net	132	103	7	47	62	(231)	98	197	16	([1])
Total	349	285	275	554	535	341	718	731	686	777
Foreign income	586	595	594	615	694	596	680	851	819	1,520
Foreign new debt and equity	34	79	70	90	217	115	328	201	[2](180)	175
Total	620	674	664	805	911	711	1,008	1,052	639	1,695
Foreign income and foreign new debt and equity in excess of expenditures less depreciation	271	389	389	251	379	370	290	321	(47)	[3] 918

[1] Data not available at this time; it will be submitted later if the committee wishes.

[2] Foreign borrowings to enable the corporation to comply with the Department of Commerce's OFDI regulations during 1972, were not completed until the 1st 60 days of 1973, as permitted by the regulations.

[3] Preliminary subject to data on other foreign expenditures, net.

TABLE 1
Oil Production by Selected Firms, 1975

	crude and natural gas liquids; 000's of barrels per day (mb/d)	share of overall production
Imperial	265	16.33%
Texaco	139	8.56
Gulf Oil	131	8.07
Mobil Oil	113	6.96
Chevron Standard	110.6	6.81
Amoco Canada	98.1	6.04
Shell Canada	81.9	5.04
Hudson's Bay Oil & Gas	78	4.80
Aquitaine Co.	68.2	4.20
Sun Oil*	56.7	3.49
	1141.5	70.30
All other producers	481.6	29.70
TOTAL PRODUCTION	1623.1	100.00

SOURCES: *Oilweek*, May 17, 1976; National Energy Board *Annual Report*, 1975. The list of Canada's top producers is published regularly in *Oilweek*. Share of production was calculated by using the NEB's estimate of total Canadian output for the year. The NEB's estimate of oil production is higher than *Oilweek*'s.
*Includes output from Great Canadian Oil Sands.

the large companies. Indeed, the overwhelming share of Canadian gas production is controlled by the oil industry and therefore, in analyzing Canadian natural gas policy, it is not only relevant but indeed essential to understand the structure, attitudes and influence of the oil industry.

According to *An Energy Policy for Canada*, published by the Federal Government in 1973, the Canadian oil industry

> . . . is dominated by foreign-controlled firms which account for over 91% of the assets and 95% of the sales of the industry. Approximately four-fifths of the non-resident controlled assets are held by U.S. residents or U.S. resident controlled corporations. Of the firms which have operations in all phases of the industry—from exploration to marketing—non-resident control is 100%. This group of "integrated companies" accounts for $5.5 billion or 51% of the total industry assets.[11]

TABLE 2
Gas Production by Selected Firms, 1975

	Gross natural gas production; millions of cubic feet per day (MMcf/d)	share of overall production
Shell Canada	621	9.31%
Pacific Petroleum	480	7.19
Gulf Oil	470	7.04
Amoco	426	6.38
Hudson's Bay Oil & Gas	419	6.28
Imperial Oil	412	6.17
Mobil Oil	356	5.33
Aquitaine Co.	335	5.02
Chevron Standard	263	3.94
PanCanadian Petroleum	241	3.61
	4023	60.27
All other producers	2648	39.73
TOTAL PRODUCTION	6671	100.00

SOURCES: *Oilweek*, May 17, 1976; National Energy Board *Annual Report*, 1975.

The domination of the world oil industry by seven giants—Exxon, Royal Dutch/Shell, Gulf Oil, British Petroleum, Texaco, Mobil Oil and Standard Oil of California—is therefore reflected in the oligopolistic nature of the Canadian oil industry.

A widely-used index of the market power of individual firms in an industry is the share of overall production which each company controls. Statistics for Canadian oil and gas production, by company, are reproduced in Tables 1 and 2.

In 1975, Canada's top ten producers accounted for over 70 per cent of aggregate crude oil and natural gas liquids production. This represents an appreciable increase in the concentration of production among the largest firms in the five years since 1971 when the top ten producers controlled only 58 per cent of output.[12] With the exception of Amoco, which is a subsidiary of the Standard Oil Company of Indiana, six of the first seven producers are affiliates of the major multinational oil corporations. No Canadian-controlled company appears on the list.

Production concentration is not as striking in the case of the gas industry. Nevertheless, the top companies here control over 60 per cent of total output.

As can be readily discerned from Tables 1 and 2 the overlap between Canada's top oil and gas producers is almost total although the relative ranking of these companies varies somewhat in both fuels. The five subsidiaries of the oil 'majors' listed in both tables—Imperial, Shell, Gulf, Mobil, and Chevron *— account for 43.2 per cent and 31.8 per cent of oil and gas production respectively in Canada today. This share of overall output can be expected to increase further with the development of Mackenzie Delta reserves.[13] Only one Canadian-owned company, Pan-Canadian Petroleum, ranks among the top domestic gas producers.

The structure of the Canadian oil and gas producing industry is characterized by the concentration of production among a few foreign-owned firms, most of which are subsidiaries of the international 'majors'. The influence which these companies can bring to bear is directly related to their market power and represents therefore a formidable factor in the Canadian policy process. Before examining specific instances of that influence, let us review the values which underline the thrust of these companies' decisions.

2 VALUES

By definition, the Canadian oil and gas producing industry is continentalist in outlook; not only is it overwhelmingly owned by American corporations from which it often borrows technology, capital and personnel, but, more importantly, it operates according to economic criteria, such as the maximization of profits, which *a priori* dictate a disregard for "the imaginary border"[14] between Canada and the United States. Thus, the behaviour of Canadian-owned corporations such as Panarctic Oils Ltd., which is 45 per cent government-controlled, does not differ materially from that of a wholly-owned subsidiary of a foreign corporation.[15] Indeed, Panarctic's President, Dr. C. Hetherington, prides himself on the fact that the company to date "has been run largely as a business" and recommends that a Polar Gas pipeline be built as soon as possible—even if its entire throughput were exported to the United States—because

> we have sunk an awful lot of money into exploring the Arctic Islands and we would like to start getting some of this money back, so that we can continue exploration on a larger scale.[16]

---

*A figure for Texaco's gas production was not available.

In the past the Canadian oil industry has not hesitated to call for continentalist oil and gas policies. In 1970 J.A. Armstrong, then the President of Imperial Oil, declared that "a Canadian commitment to a continental approach is needed so that the industry can begin to formulate long term planning."[17] If the petroleum industry today has understandably muted its public advocacy of continentalism, it has done so not because it now repudiates this concept, but rather because it has adopted a new strategy which seeks to desensitize the issue by transforming it into an essentially technical problem of market definition, export formulas, exploration incentives, etc.

It is instructive to examine the arguments which the petroleum industry made at the National Energy Board hearings on natural gas supply and demand held in late 1974 and early 1975. The arguments illustrate quite clearly this more sophisticated strategy. The producers made a number of recommendations at the hearings; among these were the early construction of a Mackenzie pipeline, a substantial increase in prices, the honouring of existing export licenses, a relaxation in the National Energy Board's export protection formula, the granting of additional incentives to develop supplies and the encouragement of energy conservation. All of these recommendations can be reduced to two fundamental principles: the commodity pricing of Canadian gas to reflect world oil prices; and the provision of early market access to all discovered reserves.

In its oral testimony at the hearings, Imperial Oil argued that

> gas prices should be allowed to rise through market forces to a level competitive with the international price of oil products. This would induce conservation, would encourage the use of substitute fuels, and would limit further conversion from other fuels to gas, all acting to dampen the growth in demand. It would also provide incentive to increase gas supply, provided the producer shares in the benefits of the increased price.[18]

In advocating that Canadian gas prices be determined by world prices, Imperial Oil and the other gas producers are ostensibly arguing that reliance on free market forces, the 'invisible hand', will result in the optimum allocation of resources.

This argument, appealing as it may be in a capitalist-oriented society, is couched in a language which does not make clear the real issues involved. Without debating whether the precepts of neoclassical competitive theory have ever applied to the transactions of the oil and gas industry, it is clear that the acceptance of world oil prices as the determinants of domestic oil and gas prices would, in effect, tie an important aspect of Canadian natural gas policy to the vagaries of international geopolitics and therefore abrogate this country's jurisdiction over that component of its energy policy. By the same token, it would also create a substantial profit potential for gas producers as present production costs in the western provinces average only $.08 per thousand cubic feet (Mcf) while costs in the Mackenzie Delta are projected by the Canadian Petroleum Association, perhaps conservatively, at between $.13 and $.25/ Mcf.[19] By contrast, under commodity pricing, gas prices would reach over $2/Mcf in the near term.[20]

The chief argument which the gas producers advance in favour of equating the public good with their private interest is that commodity pricing will induce an increase in supply. Higher prices, the companies argue, translate themselves into higher earnings and, therefore, larger exploration budgets. This argument, however, does not appear valid in the case of at least one oil company, Imperial, which, despite a substantial increase in earnings between 1972 and 1975, allocated the same funds to exploration in those years.

The figures which the producers presented at the NEB gas hearings to support their contention that higher prices would stimulate supply tended to indicate that although higher prices may lead to increased supply in the long term, in the short term, supply is highly price inelastic. Indeed, according to the Canadian Petroleum Association, a 400 per cent increase in price would yield as little as a 5 per cent increase in production by 1980, at a marginal cost estimated at $42/Mcf.[21] Thus, in the short term at least when increased supplies would be most needed, a rise in price appears quite ineffective in enticing higher production levels.

In the long term, however, the opposite problem may be created as commodity pricing, according to industry forecasts, results in substantial excess production capacity. The producers' recommendation that prices be raised immediately to commodity level is tantamount to asking the Canadian consumer to pay high prices in the seventies in order to generate exportable surpluses in the eighties.[22]

It has been calculated that if the price levels for natural gas forecast by the CPA were achieved, Canadians would pay an extra $150 billion during the next twenty years over and above current prices and the normal rate of inflation.[23] Although the macroeconomic impact of an increase of such magnitude may be difficult to estimate, at least two additional cost areas can be identified: the establishment of petrochemical plants in Alberta may be prevented without extensive government subsidy;[24] Canadian companies using natural gas could be placed at a competitive disadvantage with American firms buying gas at the prices regulated by the Federal Power Commission.

The commodity pricing argument cannot be dissociated from the second principle which the oil companies enunciated at the hearings, namely the early market access for all discovered gas reserves. Both constitute facets of the same coin and were presented as logical concomitants of each other.

Stated simply, early market access calls for reserves to be produced as they are discovered; any surpluses which might arise as a result would be exported. Carried to its extreme, the argument suggests that in the admittedly unlikely event that all of Canada's remaining undiscovered potential were found in the next decade, it would be immediately exploited.

Indeed, the very development of reserves for domestic needs may sometimes depend on an export component. Imperial Oil, for example, stated in its brief to the gas hearings that

> the extremely long lead times for exploration, development, and provision of the required transportation systems represent a significant delay in the perceived cash flow generation. Added delay associated with waiting for growth in the domestic market may render exploration uneconomic.[25]

According to this logic, exports are the *sine qua non* to the satisfaction of Canadian demand. Although the gas producers claim that early market access would reduce the inventory costs imposed by shut-in gas reserves and thus increase the cash flow available for exploration,[26] it is self-evident that the finite nature of natural gas reserves constitutes a limit beyond which no amount of exploration will be successful. The argument that domestic supplies would in effect be 'created' as a result of a vigorous export policy is one which is therefore demonstrably unsound in the long term and extremely suspect in the short term as well.[27]

This argument, it should be noted, is marked by a curious circular form of reasoning: the development of domestic supplies is made dependent on export volumes and both result from higher prices; higher prices, however, are made neccessary precisely because the initial export has depleted the least-cost reserves. Thus, ". . . the volumes of gas required to justify the export of gas are seen to depend on the export of gas for their development."[28] Hence, although the producers invariably reccommended throughout the gas hearings that Canadian needs be met before exports were approved, the policies they advocated precluded the securement of this very objective.

Another consequence of exporting large volumes of natural gas is that Canada would whittle away its comparative advantage in this energy source by unnecessarily accelerating the depletion of a premium fuel for which substitutes of comparative quality have not yet been developed in many end uses.[29]

Early market access is but another way of stating that artificial barriers to trade, such as export formulas, should be removed and that continental market demand should dictate the rate of resource development. Under such a system, inter-company agreements would be allowed to govern the production and sale of natural gas. Canadian gas policy would be reduced effectively to the aggregate of contracts between sellers and purchasers.

It is evident that if the style of the pronouncements emanating from oil and gas industry officials has been modified recently in response to a growing Canadian nationalism, the substance of the message remains the same. Acceptance of the tandem arguments of commodity pricing at international levels and early market access would inevitably lead to greater continental integration. Not only would the development of an alternative industrial strategy based on a two price system be obviated but Canada's role as a resource hinterland would be reinforced by the increased levels of its raw materials exports.

We have seen that the dominating heights of the Canadian natural gas industry are occupied by the producers whose affiliation to the multinational oil companies has given them a substantial degree of power; we have also seen that these producers continue to favour a continentalist energy policy. It is not surprising therefore that this combination of power and values has given Canadian gas policy a strong continentalist bent.

3 INFLUENCE

The oil and gas industry has been remarkably successful in imprinting its stamp on Canadian energy policy. This influence derived partly from the self-righteous indifference towards energy policy which the federal government has shown until recently but also from the industry's near monopoly on expertise and information. Indeed, as late as 1974, the then Minister of Energy, Mines and Resources, Donald Macdonald, was forced to admit in an *Oilweek* interview that "one of the difficulties facing the Canadian government is that it is virtually dependent on the major international companies for its sources of information."[30]

It is ironic therefore that one of the means chosen by the federal government to gain a clearer understanding of the energy situation has been to ask the National Energy Board to conduct public hearings on the supply of and demand for crude oil and natural gas. As the main intervenors in both these hearings have been the major oil companies to which Mr. Macdonald referred, one may infer that government policymakers will continue to rely on industry-supplied information in the near future in formulating policy.

Despite the growing assertiveness of both levels of government, it is probable that the oil industry will remain one of the principal sources of influence in the development of Canadian natural gas policy. In order to gain an appreciation of the possible weight of that influence in the future, it may be useful to review some examples of the industry's success to date in furthering its own interests.

4 THE NATIONAL OIL POLICY 1961

The National Oil Policy established a protected market for Canadian crude west of the Ottawa Valley to allow the domestic oil industry to grow free from the competition imposed by the importation of cheaper offshore oil. As natural gas and fuel oil can be substituted for each other in many end uses, a direct consequence of the Oil Policy was to curtail the penetration of western gas into eastern Canada as it could not compete with imported oil.

The National Oil Policy has therefore played an important role in structuring Canadian gas policy: by limiting the size of the domestic market, it created a greater potential for export. It has recently been acknowledged that Quebec's dependency on oil for a disproportionate share of its energy budget will impose a significant burden on Canada's balance of payments now that Canada has lost its net self-sufficiency in petroleum.

The National Oil Policy formed the cornerstone of Canadian energy policy throughout the sixties. There exists some evidence to suggest that the oil industry played a decisive role in the formulation of that policy.[31]

In the late fifties, the inclusion of the large Montreal market into the western half of the country served by domestic oil had been seriously considered.[32] This move, however, had been vigorously opposed by the major oil companies for two main reasons: they alleged that a downward pressure on wellhead prices would be exerted by the high transportation costs necessary to bring western oil to Montreal; secondly, they argued that

> . . . if Canadian crudes were marketed in Montreal under these circumstances, it might not be possible for Canadian crude prices to continue to be based on the price of Illinois crude at Sarnia.[33]

This second argument explains why the major oil companies favoured the protection of the domestic market no further east than the Ottawa Valley Line. Until the federal government imposed price controls on Canadian crude in September 1973, Canadian oil prices had been determined in the Chicago market.[34] The Canadian majors feared that if they had to serve the Montreal market, they would lose their export markets—as western production capability could not accomodate both—and thus their only justification for using the American pricing system which was extremely profitable as domestic oil production costs have traditionally been substantially lower than similar costs in the United States.[35] Indeed, the adoption of the American pricing system leads one to suspect that Canadian oil has always been priced at an artificially high level and that Canadan oil producers have consequently been reaping super- normal profits.

The reasons why the National Oil Policy Line was drawn along the Ottawa Valley rather than further east cannot be accounted for exclusively by the pressures brought to bear by the Canadian affiliates of the American majors. The Venezuelan subsidiaries of the same majors were also opposed to extending the InterProvincial PipeLine to Montreal as it would have meant the loss of an important market. These companies found an ally in the American government which feared that such a reduction might have affected Venezuela's political stability; the United States therefore agreed to allow an increase in Canadian oil imports if the Montreal market continued to rely on Venezuelan crude.[36]

The Canadian government, which was concerned about the depression in the domestic oil industry resulting from declining markets,[37] chose to encourage oil sales to the United States rather than provide access for domestic crude to eastern Canada. This policy corresponded closely with the interests of the American majors as, under the auspices of the National Oil Policy

> . . . the industry itself . . . found new markets for Canadian oil in the U.S. . . . and . . . saved the Montreal market for imported oil, thereby preventing the development of substantial pressure on U.S. east coast markets from Venezuela crude oil.[38]

The National Oil Policy constitutes an outstanding example of the susceptibility of government policies to manipulation by the major oil companies; having first priced Canadian oil at an artificially high level which precluded its competitiveness in Montreal, the majors then exerted pressure on the American government to permit the importation of some Canadian oil, for ostensibly strategic reasons, in order to prevent the inclusion of Montreal into the protected Canadian market. "The consequence," as Milton Moore, an economist at the University of British Columbia, wrote, "is that Canadians east of the Ottawa Valley have paid monopoly prices while Canadians west of that line paid even higher prices."[39]

The adoption of the National Oil Policy in 1961 must be seen as a major step towards a greater integration between the Canadian and American energy economies. By erecting a barrier at the Ottawa Valley for domestic oil and gas, the Canadian Government created in effect an oil and gas surplus which was available for export to the United States. The National Energy Board export recommendations throughout the sixties were made possible to a large extent by this arbitrary division of the Canadian energy market.

The sobering realization that Canada's potential petroleum-bearing basins would perhaps not be as prolific as had been initially estimated was surprisingly slow to develop in government circles. An analysis of the reasons for the longevity of the myth that Canada harboured a vast treasure house of energy resources provides a valuable clue in understanding the complacency which until so recently characterized Canadian energy policy.

This myth was actively promoted by the oil industry through the use of highly misleading, if not unsound, methods of potential reserve calculations. The volumetric method of estimating potential hydrocarbon reserves in a sedimentary basin is still an important technique in use today. Using available geological data, an estimate is reached of the volume of sedimentary rocks in a basin; this volume is then multiplied by a recovery factor which is a figure, based on comparable basins, of the number of barrels of oil that may, on an average, be ultimately recovered from a cubic mile of sediment. In reaching the deceptively precise figure of 120.9 billion barrels of oil and 724.8 trillion cubic feet (Tcf) of gas as the potential hydrocarbon reserve in Canada, the Canadian Petroleum Association (CPA) assigned the value of 46,000 barrels of oil per cubic mile of sediment and further assumed that 6000 cubic feet of gas would be associated for every barrel of oil.[40]

The clearest indication of the unreliability of the volumetric method used by the CPA in calculating the extent of Canada's potential gas reserves in 1969 is perhaps best illustrated by the substantial downward

TABLE 3
Potential Gas Reserve Estimates

Expert	Year	Quantity (Tcf)
Canadian Petroleum Association	1969	724.8
Geological Survey of Canada	1972	906.2
Geological Survey of Canada	1973	782.9
Canadian Society of Petroleum Geologists	1973	577.5
Canadian Petroleum Association	1974	750 - 300
Gulf Oil Canada Ltd.	1974	537 - 435
Shell Canada Ltd.	1974	440
JLJ Exploration Consultants Ltd.	1974	423
Geological Survey of Canada	1976	282

SOURCES: The first four estimates are taken from Volume 1, *An Energy Policy for Canada*, p. 91. The following four estimates are contained in the submissions to the National Energy Board gas hearings by the CPA, Gulf, Shell and Foothills Pipe Lines Ltd., respectively. The last estimate, taken from *An Energy Policy for Canada*, p. 62, was prepared using probability theory and assumes a fifty per cent probability that as much gas will be found.

revisions of various estimates which have been made recently. These revisions are listed in Table 3.

Despite the highly questionable value of potential reserve estimates, these have undoubtedly had important policy implications.[41] Dr. Ken North, a geologist at Carleton University, stated in 1972 how

> every member of the federal cabinet, every provincial premier, scores of oil company spokesmen, and thousands of journalists, broadcasters and members of the investing public are totally convinced that these "reserves" are actually known to exist; they merely have to be located.[42]

Not only were these estimates accepted without reservation, but indeed, the view that both crude oil and natural gas might soon become obsolete and thus ought to be exploited as rapidly as possible was also one which coloured government policy.

The fallacious belief that Canada's proven depleting reserves would be replaced by much larger ones in the near future created a climate which was conducive to high export levels. Eric Kierans claims, for example, that the potential reserve estimate submitted by the CPA during the 1970

National Energy Board gas export hearings ". . . was the real basis for the Cabinet decision to permit the export of 6.4 Tcf."[43] Over the years, Canada has signed long term contracts for the export of 22.7 Tcf of natural gas to the United States, 11.8 of which remained to be exported at the end of 1976;[44] in 1974, thirty-nine per cent of Canada's marketable gas production was exported.[45]

Canada's past export policy, premised on industry-supplied information indicating the existence of a vast reserve potential, has led to the depletion of the cheapest, most accessible reserves, reserves which will have to be replaced by harder-to-find, more expensive gas in both the conventional and the frontier areas. Here too, the oil industry has succeeded in advancing its own interest, thereby influencing the course of Canadian natural gas policy. As with the National Oil Policy, this success has led to a closer integration between Canada and the United States.

The National Oil Policy and the mystification surrounding Canada's resource base do not constitute, of course, the only examples of the industry's influence on public policy although they certainly rank among the most important ones. Many other cases have been successfully documented: Andrew Thompson and Michael Crommelin state, for instance, that the federal government gave the oil industry *carte blanche* to draft the Canadian Oil and Gas Regulations of 1961 which, until they were revoked in 1970, applied to all federally-controlled lands. "As a result," they write, "we have a resource 'give-away' unparalleled in any country in modern times."[46] It is not yet clear whether this 'give-away will be precluded under new regulations whose general terms were outlined in May 1976 by the Minister of Indian and Northern Affairs, Mr. Buchanan.[47]

Another instance of the industry's ability to influence government policy was provided in August 1966 when the Canadian Cabinet, rejecting a recommendation from the National Energy Board, refused TransCanada PipeLine's application to build a large-diameter line going through American territory (The Great Lakes Transmission Co.) The government was only reaffirming its support for a policy which it had enunciated ten years earlier by stipulating that TransCanada PipeLine be built entirely on Canadian territory. However, following "a high level meeting . . . between senior Cabinet ministers and one of the most formidable groups of Canadian businessmen ever assembled . . ."[48] the Cabinet overturned its previous decision and accepted the Great Lakes proposal with only minor modifications.[49]

The industry's most widely-used weapon remains the threat to curtail exploration. This threat has been applied often-times in the past: when the Alberta Conservation Board refused an application to export gas

The Lougheed

photograph by Ken Bell.

Peter Lougheed
Premier of Alberta,
December 4, 1973:

The magnitude of the Alberta oil sands is so great it can supply oil surplus to Canadian needs for export to the United States. Alberta would like a Canadian-United States treaty by which . . . Alberta's oil would be sold in the U.S.

Brothers

Imperial Oil Ltd.

D.D. Lougheed
Senior Vice-President and Director,
Imperial Oil Ltd.,
June 9, 1976:

When thinking about the various pipeline proposals, the old saying that 'a bird in the hand is worth two in the bush' is most appropriate. The Arctic Gas project is the bird in the hand because it is solidly backed by the very large, proven reserves of gas at Prudhoe Bay. We Canadians have a chance now to take advantage of this to get our own gas to market quickly by riding piggyback. The Maple Leaf project and the Alaska Highway proposal are the birds in the bush.

from the province in 1951, Gulf threatened to abandon its exploration and development plans worth over a third of a billion dollars.[50] In 1967, Imperial Oil announced the suspension of a major exploration programme until the Carter Commission tax proposals were dealt with by government.[51] In 1974, as a result of the Federal-Provincial confrontation over resource revenue, several companies announced substantial cuts in their budgets in a series of well-publicized moves.[52] Although the industry claimed it was saddled with an unrealistic tax burden as a consequence of the non-deductibility of royalty payments for corporate income tax purposes, gross production revenues nevertheless exceeded expenditures by 1.6 billion in 1975.[53]

Thus, although there may have been grounds for having doubted the severity of the industry's claimed plight in the mid-seventies, the oil and gas explorers demonstrated collectively that they had the power to exert a direct impact on the rate of reserve addition, and hence future gas supply, by reducing their exploration and development programmes.

6 CONCLUSION

The Canadian Natural Gas industry, despite a substantial degree of Canadian ownership concentrated in the transmission and distribution sectors, is nevertheless dominated and guided by the major international oil companies. These companies' control over a disproportionate share of Canadian gas production has allowed them to set the pace and tone of development and thus the direction of Canadian gas policy. In a climate of government non-interference fostered by the free market ethic, the oil and gas producers were successful in arrogating to themselves a formidable power which they wielded to impose their values. Their leverage in swaying policy was reinforced by their perceived indispensability and by the government's sharing of similar ends, such as growth.[54]

The oil industry's success in moulding Canada's energy policy can be observed in the degree of Canadian-American integration in the energy field today. Leonard Waverman writes:

> The degree of interdependence in Canadian-American energy relationships is large and complex—Canada exports half its natural gas and oil production to the U.S. while importing half its coal requirements from the U.S. American companies own a large share of Canadian oil and gas reserves, pipelines and refining capacity. Together, the two countries have planned a large hydro electric system—the Columbia River Project. Amid these relationships, is calling for a continental energy policy merely asking for a formal label?[55]

Perhaps a more relevant question to ask at this time is whether the oil and gas industry will be as successful in the future in influencing the direction of Canadian energy policy. The number and significance of the changes which have occurred recently concerning energy policy, the most important of which is the more assertive presence of both senior levels of government, may lead one to forecast a much reduced level of industry influence in the formation of natural gas policy. Although it is indeed likely that the industry will lose its traditional autonomy, there is a natural if fallacious tendency to overstate the importance of recent events in this regard because of the dramatic circumstances which accompanied them (viz. Federal-Provincial confrontation over royalties, Alberta-Ontario confrontation over petrochemicals and prices, Canadian-American confrontation over export prices).

It is important to note that the oil and gas industry, however circumscribed its role may become as an outcome of these events, has not seen its traditional function as an energy supplier challenged. Furthermore, it is not clear whether its regulation to an apparently secondary position is the result of conscious political efforts to restrict the latitude of free enterprise in this domain or simply a fortuitous consequence of a Federal-Provincial conflict aimed at determining dominance in the resource field. The creation of Petro-Canada, for example, may not constitute a fundamental challenge to the industry as much as a catalyst to promote exploration and development in the frontier areas. Indeed, at his first press conference as Petro-Can's President, Mr. Maurice Strong stated that the company's prime role is to supplement private industry, not displace it.[56]

The present thrust of government policy towards self-reliance and ultimately self-sufficiency in hydrocarbons, as embodied in *An Energy Strategy for Canada*, will afford the oil and gas industry a substantial scope for influencing policy. The oil industry has actively supported the principle behind self-sufficiency for a long time as it implies a very large potential for growth.[57] Not surprisingly, the industry's support of conservation, whose contribution to self-sufficiency could be very significant,[58] has been much more muted.

For the oil and gas industry, the adoption of the policies outlined in *An Energy Strategy for Canada* could include government financing assistance for pipeline construction[59] and a much greater cash flow to raise the $90 billion needed to meet capital requirements by 1990.[60] The government's recognition that "the degree to which funds can be generated internally through higher prices will be critical"[61] implies a readiness to consider a lower tax regime as well as higher prices to assist the industry.

That the pursuit of a policy of self-sufficiency in hydrocarbons, particularly oil, constitutes a policy objective for Canada which entails a substantial economic risk[62] bears testimony either to the industry's ability to influence policy in its own interest or to the government's perception of the congruence between the national purpose and private interests.

The future of the Canadian natural gas industry does not appear at this time therefore to be threatened by greater government intervention. If recent experience provides any indication, the industry stands to benefit considerably: government intervention has so far resulted in higher prices and the preservation of exploration activity in the high arctic. In *Oilweek's* words,

> Canada's natural gas industry, hampered for more than two years by political and economic uncertainty, appears to have emerged from this period of gloom into the bright sunshine of spiraling activity brought about by higher prices and an improved reserves picture.[63]

1 Natural gas cannot be transported or marketed with the same facility as crude oil or its products and requires a costly investment in pipelines in order to deliver it to the consumer. The size of the investment and the need to amortize the pipeline over a period of time dictate secure markets without competition from other gas distributors. In order to protect the consumer against the potential abuse of monopoly power, governments have traditionally instituted public boards to regulate the activities of both transmission and marketing companies.

2 Norcen Energy Resources Ltd.: *Annual Report,* 1975. Norcen also markets liquified petroleum gas products (propane and butane) through Cigas Products Ltd. and holds interests in coal exploration and distribution as a result of an 81 per cent ownership of Coleman Collieries.

3 William Kilbourn: *Pipeline,* Toronto: Clarke, Irwin and Company, 1970, 98.

4 *The Globe and Mail,* April 22, 1975.

5 PanCanadian Petroleum Ltd.: *Annual Report,* 1975.

6 Financial Post: *Survey of Oils 1976,* Maclean-Hunter Ltd., 1976, 92.

7 See Ian McDougall: "The National Energy Board: Economic Jurisprudence in the National Interest or Symbolic Reassurance?" in the *Alberta Law Review,* Vol. IX, No. 2, July 19, 1973, pp. 351, 353 and discussion *passim* 347 to 357.

8 The following is a list of companies which at one time or another have belonged to the Arctic Gas consortium:

The Alberta Gas Trunk Line Co. Ltd.
Alberta Natural Gas Co. Ltd.
Atlantic Richfield Co.
Canada Development Corp.
Canadian National Railways Co.
Canadian Pacific Investments Ltd.
Canadian Superior Oil Ltd.
Colorado Interstate Corp.
The Columbia Gas Transmission Corp.
The Consumer's Gas Co.
Exxon Co., USA
Gulf Oil Canada Ltd.
Imperial Oil Ltd.
Michigan-Wisconsin Pipe Line Co.
Natural Gas Pipeline Co. of America
Northern and Central Gas Corp. Ltd.
Northern Natural Gas Co.
Numac Oil and Gas Ltd.
Pacific Lighting Gas Development Co.
Panhandle Eastern Pipeline Co.
Pembina Pipe Line Ltd.
Shell Canada Ltd.
Standard Oil Co. of Ohio
Sunoco E&P Ltd.
Texas Eastern Transmission Corp.
TransCanada PipeLines Ltd.
Union Gas Ltd.

9 For a discussion of the effect of interlocking directorships on corporate economic power, refer to John Porter, *The Vertical Mosaic*, Toronto: University of Toronto Press, 1965, particularly Chapter XVII. In Appendix XI of Wallace Clement's *The Canadian Corporate Elite*, Toronto: McCelland and Stewart, 1975, the reader will find a list of interlocking directorships among the corporations Clement lists as dominant.

10 This was the conclusion of Mr. L.M. Rasmussen, the then newly elected President of the Canadian Gas Association in a *Globe and Mail* interview, May 29, 1975.

11 Department of Energy, Mines and Resources: *An Energy Policy for Canada—Phase I*, Ottawa: Information Canada, 1973, Vol. 1, 239, 240.

12 *Ibid*, Vol. 1, p. 250.

13 Oil discoveries in the Mackenzie Delta have been made by Imperial Oil and Shell; gas discoveries by Imperial Oil, Shell and Gulf-Mobil.

14 In 1969, in Washington, J.J. Greene, then the Minister of Energy, Mines and Resources, declared that he favoured a continental energy policy so that "the people will benefit, and both countries will benefit, irrespective of where the imaginary border goes." Quoted in James Laxer, *The Energy Poker Game*, Toronto: New Press, 1970.

15 Porter, *op. cit.*, p. 270, notes that the behaviour of Canadian and American firms operating in Canada is quite similar, a comment echoed in "Resources: Implications of Ownership" by Arthur Cordell in *Essays on Aspects of Resource Policy*, A Background Study for the Science Council of Canada, May 1973, Special Study No. 27, 105.

For a critique of Panarctic, see Pat Deutscher, "Panarctic Oils Limited: The government and the Energy Business," in *Canadian Forum*, June-July, 1973.

16 *IN THE MATTER OF the National Energy Board Act; and IN THE MATTER OF an inquiry, hearing and determination of the reserves, requirements and deliverability of Canadian Natural Gas in relation to reasonably foreseeable requirements for use in Canada and potential of export. ORDER NO. GHR-1-74, hereinafter referred to as the National Energy Board Gas Hearings.* Transcript, pp. 553 and 549. It will be interesting to see in the years to come whether Panarctic's parent, Petro-Canada, already a significant presence on the Canadian oil scene through its acquisition of Arcan, its participation in Syncrude and its sizeable exploration programme, will also succumb to the same pressures.

17 Quoted in Laxer, *op. cit.*, p. 12. Mr. Armstrong is now Chairman of Imperial Oil.

18 Imperial Oil Ltd.: "Opening Statement" at the National Energy Board gas hearings, p. 3.

19 Sherman H. Clark Associates: *Canadian Natural Gas Supply and Requirements: 1973-1995.* Prepared for the Canadian Petroleum Association for presentation to the National Energy Board Gas Hearings, pp. 194 and 229.

20 On a Btu equivalent basis, one Mcf of gas is approximately equal to one sixth the energy contained in a barrel of oil. International oil prices average about $11.50/bbl at the time of writing and landed in Toronto would be close to $13/bbl. The difference between production costs and price is, of course, not all profit as taxes and royalty payments would have to be included.

21 Industrial Gas Users Association: "Final Argument" to the National Energy Board Gas Hearings, p. 2.

22 In an article in Northern Perspectives (published by the Canadian Arctic Resources Committee, Vol. III, No. 2, 1975), Bruce Willson argues along similar lines.

23 Bruce Willson, letter to *The Globe and Mail,* Feb. 28, 1975.

24 *The Globe and Mail,* March 1, 1975.

25 Imperial Oil Ltd.: *Submission to the National Energy Board Gas Hearings,* p. 13.

26 The gas producers point to a correlation between high export levels and high rates of discovery during the sixties as establishing the soundness of their argument. It is not clear, however, whether a significant casual link does exist between gas exports and reserve additions during that decade. Indeed, as gas is often found as a byproduct of oil exploration, it is conceivable that the creation of a guaranteed market for oil through the implementation of the 1961 National Oil Policy contributed more indirectly to the growth in gas reserves than export revenues *per se,* particularly as gas export prices remained low during the entire decade.

27 Dome Petroleum, which incidentally is a Canadian company, argued in the Calgary phase of the National Energy Board Gas Hearings that:

> . . . the already connected Canadian customers would be far better protected by the addition of a short term export market than they would by the addition of additional market areas in Canada which became dependent upon a gas supply for a long term.

(Oral Statement, p. 3) This particular logic is illustrative of the reasoning which is often found in the oil industry, a reasoning which purports to demonstrate that the Canadian consumer would be better off by exporting than by increasing his consumption. That a Canadian company would hold such views is indicative of the fact that ownership *per se* does not affect the behaviour of oil companies.

28 John N. McDougall: "Regulation versus Politics: the National Energy Board and the Mackenzie Valley Pipeline" in Axline, Hyndman, Lyon and Molot, (eds.), *Continental Community?* Toronto: McClelland and Stewart, 1974, p. 253. See also Ian McDougall, *op. cit.,* p. 336.

29 Rolfe R. Colpitts "Energy Markets and their Dependence on Price: Gas Markets," in *The Royal Society of Canada's Symposium on Energy Resources* (Oct. 13 to 15, 1973) has in Appendix I (p. 135) a list of the industrial processes where natural gas is superior to fuel oil as an energy source.

30 *Oilweek,* Jan. 21, 1974.

31 It has been repeatedly alleged that an instrument of that influence was a prominent New York oil consultant, Walter J. Levy. Described by *The New York Times* as "an advisor to most of the major oil companies, most of the important consuming countries and many of the large producing countries," Mr. Levy was retained as a consultant for the Royal Commission on Energy (the Borden Commission) which recommended the implementation of the 1961 National Oil Policy. See article by Larry Pratt in *The Last Post*, Jan. 1975, Vol. 4, No. 4 .

32 A group of independent Canadian oil companies led by Home Oil had considered building the pipeline extension to Montreal and prepared feasibility studies which were presented to the Royal Commission on Energy. See the Commission's *Second Report*, July 1959, p. 111.

33 *Ibid*, p. 117.

34 Department of Energy, Mines and Resources: *op. cit.*, Vol. I, p. 66. This situation stemmed first from the foreign ownership of the Canadian oil industry which adopted the American "Gulf Coast Posted Price," an artificial pricing system imposed by the majors to stifle internal pricing competition in the United States, and also from Canada's exemption from the U.S. Mandatory Oil Import Program of 1959 on the condition that Canada not underprice its oil in the export market (i.e., the United States).

For an elaboration of the majors' oil pricing strategy during the sixties, see Francois Bregha: *Prices and Profits* presented to *Stopping the Sell-Out; the Real Energy Conference*, held in Winnipeg, March 8, 9, 10, 1974.

35 This was the case in 1957, when in a background report for the Royal Commission on Canada's Economic Prospects, Mr. John Davis wrote that ". . . finding costs have been lower and production costs much lower in western Canada than in the United States. . . "(John Davis: *Canadian Energy Prospects*, March 1957, Queen's Printer, p. 339). This was still the case in the late sixties. Reproduced below is a table from *Oilweek*, Feb. 19, 1973:

Oil costs/bbl.; average 1967-1971

	Western Canada	United States
exploration	$.30	$.27
development	.10	.29
land acquisition	.21	.21
producing facilities	.13	.19
producing costs	.27	.55
royalties	.31	.44
taxes and others	.16	.17
TOTAL	$1.48	$2.12

36 See Edward H. Shaffer: *The Oil Import Program of the United States: An Evaluation,* New York: Praeger Publishers, 1968, p. 122.

37 *Ibid.*, p. 121. As a result of the voluntary restraint imposed by the American Government on Canadian imports, Alberta's oil production declined substantially. This prompted the consideration of extending the oil pipeline to Montreal to provide new markets to western crude.

38 James D. Hilborn: "After Leduc," in James D. Hilborn, (ed.), *Dusters and Gushers,* Toronto: Pitt Publishing Co. Ltd., 1968, p. 203.

39 Milton Moore: *Problems of Optimal Management of Arctic Oil and Gas.* Paper presented to the Canadian Arctic Resources Committee's Conference on Canada's Petroleum Leasing Policy, Ottawa, March 22, 1973.

That prices in eastern Canada have historically been manipulated was established in the litigation between the Nova Scotia Power Corporation and Imperial Oil involving a $100 million suit by the company to be allowed to pass on price increases on imported oil (*Oilweek,* May 19, 1975). During the proceedings of this case, it was revealed how Exxon increased prices by transfering its oil from one subsidiary to another and how Exxon instructed its affiliates, such as Imperial, to raise prices periodically.

40 Canadian Petroleum Association: *Potential Reserves of Oil, Natural Gas and Associated Sulphur in Canada,* April, 1969, pp. 4, 6.

41 The policy implications of high potential reserve estimates were belatedly acknowledged by the CPA in its brief to the National Energy Board Gas Hearings.

Canada is not the only country to have been misled by estimates prepared by the oil industry. Figures for the remaining undiscovered oil and gas potential in the U.S. have recently been cut in half as a result of more rigorous estimate procedures. The American Government had relied upon the high estimates in the late sixties in formulating its energy policy. *The Globe and Mail,* May 8, 1975.

42 F.K. North: "Oil & Gas Resources: Delusion or Deception?" in *The Calgary Herald,* Nov. 28, 1972.

43 Eric Kierans: "The Day The Cabinet Was Misled" in *The Canadian Forum,* March 1974. Kierans was a member of the Cabinet at the time the export question was discussed.

44 The 22.7 Tcf figure was reached by adding export volumes licensed until the end of 1973 and listed in the National Energy Board's 1973 *Annual Report* (p. 63) to those committed but not yet exported and listed in the Canadian Arctic Gas Study's brief to the gas hearings on p. B108.

45 National Energy Board: *Annual Report,* 1974, p. 8.

46 Andrew Thompson and Michael Crommelin: "Canada's Petroleum Leasing Policy: A Cornucopia for Whom?" in *The Canadian Forum, op. cit.,* p. 22.

47 It is not known to what extent the industry influenced the drafting of these regulations although it appears that consultation with the industry over taxation matters has been government policy in the past. Jerry McAfee, the ex-President of Gulf Oil of Canada, for example, stated in a speech in 1974 that ". . . it used to be normal procedure for governments in Canada, when considering new public or tax policies in the energy field, to consult with the industry. . . . The policies which resulted," according to Mr. McAfee "have served Canada well." (*Financial Post,* Sept. 19, 1974).

48 William Kilbourn: *op. cit.* p. 182.

49 The reason given for this policy reversal in *An Energy Policy for Canada* appears particularly odd when contrasted with C.D. Howe's public determination of a decade earlier:

> The construction of the Great Lakes Transmission Co. line through the United States in 1967 and 1968, in addition to other significant advantages, did much towards achieving the . . . objective of security of supply for Central Canada by providing an alternate route from Western Canada. (Vol. II, p. 323)

At *prima facie,* it would seem more reasonable to argue that security of supply is decreased by the construction of a pipeline through foreign territory when an alternative under Canadian jurisdiction exists. TCPL's claim that the looping of its Northern Ontario line would be prohibitively expensive when compared to the Great Lakes project is made hollow by a National Energy Board staff study which concluded that after five years, the costs of the two alternatives would be almost the same. Barry D. Fisher: "The Role of the National Energy Board in Controlling the Export of Natural Gas from Canada," in the *Osgoode Hall Law Journal,* Vol. 9, No. 3, Dec. 1971, p. 572.

Canada's dependence on a number of pipelines crossing American territory (the InterProvincial oil pipeline, the Portland to Montreal oil pipeline and the Great Lakes gas line) for the satisfaction of domestic requirements can only curtail the range of future policy options, particularly those which might invite American retaliation, such as a reduction in energy exports.

50 Kilbourn: *op. cit.,* p. 19.

51 R F. Keith, D.W. Fisher, Colin E. de Ath, E.J. Farkas, G.R. Francis, S.C. Lerner: "Northern Development and Technology Assessment Systems: A Study of the Petroleum Development Programs in the Mackenzie Delta-Beaufort Sea Region and the Arctic Islands," *Science Council of Canada Background Study,* No. 34, Jan., 1976, p. 22.

52 Gulf Oil, for example, stated that it would reduce its projected capital investment by at least $100 million in 1975 (*Oilweek*, Dec. 2, 1974). Nevertheless, according to the *The Globe and Mail* (April 25, 1975), "despite increased taxes . . . Mr. McAfee revealed Gulf is increasing its capital spending this year to $250 million from $164 million in 1974 and is raising its annual dividend rate to $1 a share from 80 cents." It is obvious that the alleged Government squeeze on cash flow is relative.

53 *Oilweek*, Feb. 16, 1976; revenues had surpassed expenditures by $1.5 billion the previous year. *Oilweek*, Feb. 17, 1976.

54 The reader will find more examples of the oil industry's influence on public policy in the following works:

Canada: P. Sykes: *Sellout*, Edmonton: Hurtig Publishers, 1973.
J. Laxer: *Canada's Energy Crisis*, Toronto: James, Lewis and Samuel, 1974.
W. Rowland: *Fuelling Canada's Future*, Toronto: Macmillan of of Canada, 1974.

World: R. Engler: *The Politics of Oil*, Chicago: University of Chicago Press, 1961.
J. Ridgeway: *The Last Play*, New York: New American Library, 1973.

U.S.: Federal Trade Commission: *Investigation of the Petroleum Industry* (June 1973).

55 Leonard Waverman: "The Reluctant Bride: Canadian-American Energy Relations," in Erickson and Waverman, (eds.), *The Energy Question*, Toronto: University of Toronto Press, 1974, Vol. II (North America), p. 216.

56 *Oilweek*, Jan. 26, 1976.

57 Industry pronouncements almost invariably stress the importance of increasing supply and the resulting necessity to implement favourable government policy to stimulate exploration and development.

58 See *Energy Supply and Demand Forecasts-An Overview* prepared for the 1976 Canadian Nuclear Association Annual Meeting, Toronto, June 14-18, 1976 by P.M. Manders, Assistant Advisor Energy Policy, Department of Energy, Mines and Resources. Mr. Manders argues that a reduction in energy demand to two per cent annually could return Canada to a position of self-sufficiency almost immediately.

59 Department of Energy, Mines and Resources: *An Energy Strategy for Canada*, Ottawa, 1973, p. 110.

60 *Ibid*, p. 108.

61 *Ibid.*, p. 110.

62 In *An Energy Strategy for Canada,* the Federal Government estimates that $180 billion will be required over the next 15 years to develop Canadian energy supplies. As a result, the "average annual *increase* in energy-related expenditure, relative to historical requirements, would be roughly equivalent to about 50 per cent of total in 1975 investment in residential construction... (p. 106).

The share of GNP attributable to the energy sector would rise to 5.2 per cent between 1976 and 1990, fully 50 per cent higher than its historical average. (p. 106). Future energy investments can therefore be expected to exert an appreciable displacement effect in the economy.

63 *Oilweek,* Jan. 19, 1976.

CANADA
UNITED STATES

CLIFF PILKEY

Cliff Pilkey is Citizenship and Legislative Director for Canada of the United Auto Workers. He served in the Ontario Legislature as New Democratic Party MPP for Oshawa from 1967-71. He has been an Alderman for Oshawa from 1962-66 and 1972-76. Cliff Pilkey was President of the Oshawa and District Labour Council for ten years and President of the Canadian UAW Council, 1957-59.

5

Cliff Pilkey

Industrial Strategy and Canadian Energy Policy

In my union, the United Autoworkers, we have an obvious and special interest in the internal combustion engine and also in mass transit systems. Our livelihood is at stake in these matters. These specific concerns relate to broader concerns about an energy policy in general. We are concerned about the kind of expenditures that were made in Syncrude; about the Mackenzie Valley Pipeline and what that project will do to native rights and what it will do to the environment. We are concerned about maintaining a two price system in Canada as far as energy is concerned. As a trade union, we do not intend to ignore the fact that at least one-third of inflation in the past two years has been due to increases in the price of oil and gas.

Oil and gas could be the catalyst for job creation through the expansion of Canada's industrial base. How long are we going to tolerate a seven or eight per cent unemployment rate in Canada? Witnessing what can be accomplished in a country like Venezuela, in terms of the development of new secondary industry, one wonders if Canada's present government is going to wake up to the fact that we have potential to build a strong industrial base in Canada.

What is needed in this country is a conscious government policy to develop new industry using the millions of dollars of oil and gas profits to make the task possible. In Venezuela the policy is called "sembrar el petrol", "sowing oil". The policy means diverting state earnings from petroleum to the planning of new industry, generating jobs for the unemployed and increasing Venezuela's self-reliance by ending foreign ownership. While Canadians may not want to adopt the Venezeulan model, surely the principle is worth adopting.

Higher oil prices, the most important factor in recent Canadian inflation, have been justified on the grounds that they will stimulate exploration for new petroleum reserves. But higher prices do not guarantee additional exploration development. It is folly to provide additional revenue to foreign oil companies whose decision will be predicated on the sole interests of the specific company involved. What is to prevent them from simply paying out extra revenue in dividends to the parent companies in the U.S. or putting the revenue into exploration elsewhere in the world? If Canadians controlled the oil and gas industry and if we had a comprehensive policy in this country, then we could guarantee that any justified oil and gas price increase would be tied to a specific plan to develop new energy supplies to meet our future needs.

Higher oil prices have a very real effect on jobs in Canada. At the first conference of the Public Petroleum Association, Dr. Clifford Jutlah, an Ontario government economist, stated that higher oil and gas prices cost Ontario 22,000 jobs in 1974 and 16,000 jobs in 1975. He predicted a loss of a further 19,000 jobs in 1976.

The impact of petroleum price increases on inflation and on jobs raises very serious questions about the government's programme of wage and price guidelines. The labour movement has been saying for the past few months that what Canadians are dealing with are not wage and price guidelines, but really wage guidelines. The premise for the introduction of the programme was that it would slow down inflation and make Canada more competitive, thereby ensuring the country's ability to increase its potential to create jobs. In fact, wage controls have inhibited the trade union movement in collective bargaining, while at the same time oil and gas prices will be allowed to rise eventually to the world level. This is highly inflationary. To allow these price increases is in contradiction to the government's policy as enunciated through the wage and price guidelines.

Wage restraints hold down workers' purchasing power, thus reducing the domestic demand for manufactured goods. Energy price increases, of course, compound this problem, leaving a smaller share of the workers' already restrained income for the purchase of manufactured goods. This reduction in demand, in turn, will effectively reduce or curtail the expansion or building of new manufacturing facilities.

The government's policy of allowing energy price increases exposes the wage and price guideline programme. The government's energy policy exposes as myth and propoganda the claim that the wage and price guidelines were introduced to reduce inflation and to create new jobs by improving our international competitive position.

Another policy that contradicts the anti-inflation programme has been the decision to keep Canada's interest rates high as a way of attracting foreign captial into the resource sector. Higher interest rates and the attraction of foreign capital result in an over-valuing of the Canadian dollar. This affects Canada's ability to export manufactured goods. And despite the government's talk about Canada's competitive position in manufacturing, the reality is that more expensive energy for the manufacturing sectors of the economy causes Canada to lose whatever competitive advantage it had.

The government's policy of maintaining higher interest rates means higher capital costs. In turn, this undermines our international competitive position.

It should not be forgotten that the shift of our financial resources to energy investments is a shift into a highly capital intensive sector as opposed to investments in industries that are more labour intensive. Investment in energy carries with it little potential for job creation.

In 1974, the UAW presented a paper on an energy policy to the rank and file leadership of the union, at their Canadian UAW Council meeting. The delegates supported the proposition that this country needed a comprehensive energy policy. More recently, in 1976, our collective bargaining conference passed a resolution calling for public ownership of the resource industries in Canada. That was a step forward in the trade union movement. It is time we said to the Canadian people that the UAW, in particular, stands for public ownership in the resources sector.

In conclusion, the alternative to our present policies or lack of a policy is to bring the oil and gas industry under public ownership. This would effectively eliminate private, foreign ownership and control of a very valuable Canadian resource. Canada needs an energy policy that assures the provision of adequate supplies of energy at prices the Canadian consumer can afford.

Economic Impacts of Petroleum Price Increases

Oil and natural gas price increases have been most visible to Canadians in their direct impact on the price of gasoline and the price of home heating fuel.

At least as fundamental as these additions to the living costs of Canadians has been the impact of energy price increases on the nation's industrial well-being.

Since 1974, energy price increases have had significant impact on the industrial performance of Ontario, the centre of half the nation's manufacturing.

Ontario is, of course, a major importer of energy. In 1974, the province used about 2.8 quadrillion Btu's supplied by the following energy sources: 41 per cent oil; 24 per cent natural gas; 20 per cent indigenous nuclear and hydro power; and 13 per cent coal. The province's dependence on fossil fuels is easily apparent. Ontario's oil, natural gas and coal are virtually all provided through imports—oil and natural gas from western Canada and coal from the U.S. Much of the fossil fuel imports are used to produce electricity.

In papers appended to the Ontario provincial budgets in 1975 and 1976 and in studies carried out by the Ontario Ministry of Treasury, Economics and Intergovernmental Affairs, the impact of oil and gas price increases on the Ontario economy has been monitored.

A study of the effects of oil and gas pricing produced by the Ontario government in 1976 made estimates of which industries would be hardest hit by increased prices for oil and gas. The study concluded that an increase in the price of oil of $2.00 a barrel with natural gas indexed accordingly would lead to increases in the selling price of various Ontario industries' products of the following amounts: industrial chemicals 3.6 per cent; plastics and synthetics 2.2 per cent; cement 2.0 per cent; pulp and paper 1.4 per cent; glass products 1.4 per cent. The provincial study warned that such increases can have an especially damaging impact on export-oriented industries such as Ontario industrial chemicals—an industry where over 9 per cent of total producing costs are for energy compared with 7 per cent in U.S. production of the same products.

The Ontario government estimates that increases in oil and gas prices have been having a significant impact on employment in the province, particularly in manufacturing industries which employ close to one million people. In 1974 oil and gas price increases cost Ontario 22,000 potential jobs; the 1974 price changes resulted in a potential job loss carry over into 1975 of 16,000 jobs; and 1975 price increases were expected to reduce employment creation by 19,000 jobs in the subsequent twelve months.

For Ontario, increased energy costs totalled close to $1.5 billion in 1974 and 1975.

The inflationary effects of the price increases were crucial, not only for Ontario but for the nation as a whole. The Ontario government estimated that in 1974, oil and natural gas price increases led to inflation of 3.5 per cent in Canada, over one-third of the inflation experienced in that year. In 1975, counting the carry over effects of the previous year and the effects of new price and tax increases in 1975, oil and gas pushed inflation up by 4.4 per cent.

CYRIL SYMES

Cyril Symes is the NDP Member of Parliament for Sault Ste. Marie. Prior to his election to the House of Commons, he taught history at Sault Collegiate Institute. Cyril Symes is a member of the House of Commons Standing Committees on Broadcasting and Energy. He is energy critic for the NDP caucus.

6

Cyril Symes

The Role of Petro-Canada

Canadians have long felt that their natural resources have been mismanaged and given away to the United States at bargain basement prices. As long ago as 1932, Canadian Prime Minister R.B. Bennett stated:

> I believe that there is no government in Canada that does not regret today that it has parted with some of its natural resources for considerations that do not reflect the principle under which the Crown holds the natural resources for all people.

Unfortunately, despite the creation of Petro-Can, the nation's publicly owned petroleum corporation, the present government appears to have replaced whatever feeling of regret it may have had with either indifference or helplessness.

Petro-Can's creation at the beginning of 1976 came in response to both the general demand of the Canadian people for a more effective voice in their country's energy policy and also to the specific situation of the Parliament that established it.

The crucial question is whether Petro-Can is a viable alternative to companies such as Imperial Oil. Can it play a significant role in developing a Canadian oil industry with the goals of self-sufficiency, reasonable prices and independence?

To play such a role, Petro-Can must be shaped in the context of a sensible national energy policy. Canada's past and present energy policies have been formulated by the multinational oil companies who control 90 per cent of Canadian oil production and 99 per cent of refining. Canada's resources have been viewed by these corporate giants as part of a continental package, which has resulted in the Canadian government allowing them to export over half of Canadian oil production to the United States causing the oil shortage now facing Canada.

Had a proper Canadian energy policy been formulated, Canada could have been self-sufficient in oil (and still can be) with a Canadian price which could have been set irrespective of the world price.

Instead of the current energy policy of the federal Liberals which is aimed at self-reliance in energy by depending on foreign oil companies to develop Canadian resources at their prices, we need a policy of Canadian control of energy resources. And by Canadian control, I don't mean private enterprise control. Canadian control of the oil industry, to be effective in achieving the goals of self-sufficiency and reasonable prices, must mean public control; that is to say, government ownership. The industry must be treated as a public utility at all levels including exploration, development, production, distribution, refining, and marketing at the wholesale level to ensure that the benefits of Canada's non-renewable resources accrue to Canadians and their economy.

The key question is which agenda Petro-Can will be part of: the present policy dominated by the multinationals and American priorities or an alternative policy aimed at satisfying Canadian needs.

Because of its conviction that the petroleum industry should be operated as a public utility, the New Democratic Party proposed a publicly owned national petroleum corporation such as Petro-Can during the Twenty-ninth Parliament (1972-1974) in which it held the balance of power. The NDP proposed that the corporation would acquire Panarctic Oils to facilitate exploration and development in the North. It would also act as a national marketing agency, purchasing oil and gas both domestically and from foreign suppliers. It might take over the interprovincial pipelines. The corporation would thus pool costs and standardize prices. Initiatives in these areas would make the national corporation a signifigant force in the industry.

The role which the federal government has now spelled out for Petro-Can is quite different than that originally proposed by the NDP. Part of the explanation for the difference is to be found in recalling the circumstances under which Petro-Can was created.

In December 1973, the NDP, which held the balance of power, was prepared to defeat the Government and force an election on the proposed energy price increases and the lack of an energy policy to meet the coming crisis in price and supply. An eleventh hour policy statement by Prime Minister Trudeau which promised a single Canadian price below the world price, extension of the oil pipeline to Montreal, intensification of research into oil sand technology, and the establishment of a national oil company averted the defeat of the Government.

The Liberals were thus forced into establishing Petro-Can. While accepting an NDP proposal in name, in reality their eagerness to use Petro-Can as a significant tool in policy development was qualified, as has become evident during the present majority parliament.

On paper Petro-Can has been given broad ranging powers. The Board of Directors is accountable to the Minister responsible for the Crown Corporation and all decisions are subject to the approval of the Governor in Council. The Petro-Canada Act authorizes capital of $500 million with borrowing capabilities not to exceed $1 billion.

Further, Petro-Can has the authority to set up subsidiaries to carry out any economic activity consistent with the Corporation's objectives.

The powers set out in the Act follow closely those policy recommendations advocated by the NDP for a number of years. They include exploration for and development of deposits of hydrocarbons both in Canada and abroad. The Act also allows for the Corporation's involvement in production, distribution, refining and marketing of fuels.

On paper, Petro-Can looks like a company that means business. But as T.C. Douglas, energy critic for the NDP, said during the debate on the bill, "This Crown corporation is an economic tool and nothing more." It is what the Government plans to do with it that is crucial. I fear that Petro-Can as it is now being operated cannot be a threat nor even a challenge to private industry. It has become a victim of the policy of the Trudeau government and its predecessors of socializing losses and privatizing profits.

The history of some public or Crown corporations in Canada illustrates how nationalization can be used to benefit private corporations. Government takeover and formation of the CNR, for example, not only compensated private owners by assuming and repaying their debts, but enabled an unprofitable railway line to continue operating by providing government freight and passenger subsidies.

Past policies of nationalization or public ownership have left corporate structure and its power undisturbed. Those who have benefited most under these conditions are the largest users of the nationalized service or goods, usually large private corporations.

Under the deliberate policy of having Crown corporations subsidize private losses, is it any wonder that most public enterprises in Canada operate at a loss? Is it any wonder that this has reinforced business' argument to the public that governments cannot run anything?

It would appear that the international oil industry in Canada has recently discovered that there can be benefits derived from governments

Carving Out a Role for Petro-Can

MAURICE STRONG *(Chairman of Petro-Can):* We are not here to throttle or impede private industry but to add a new dimension. . . . The existence of Petro-Canada may be the best guarantee for the continued health and survival of the private industry. I'm confident that private companies will welcome Petro-Canada and, paradoxically, will welcome it as one of the best guarantees of their continued existence.

becoming involved in limited ownership and development of the energy industry. A remarkable case to remember is that since the Arab takeover of the oil fields in the Middle East, the multinational oil companies have doubled their net profits. They discovered that the conversion and marketing of oil was more profitable than ownership at the wellhead. As one oil company official explained:

> The capital investments involved in exploration and production are simply getting too huge for us to bother with. Interest rates fly up and down unpredictably. There's never any guarantee we can raise the necessary captial. So why bother? Let government explore for oil and develop it. We'll gladly pull it out for them on a fee basis and refine it and market it for them. That way we're guaranteed a profit—and the people are guaranteed they'll get oil.

If this philosophy is prevalent in Canada, it is all the more important that Petro-Can be given more than a limited role to play in developing future energy. It must not become a sophisticated method of subsidizing private industry.

What has Petro-Can been doing since its incorporation? It has undertaken a drilling programme worth $24 million off Canada's east coast on property leases controlled by Shell Oil Canada Ltd., Amoco Canada Petroleum Ltd., and Mobil Oil Canada Ltd. The extra capital contributed by Petro-Can is welcomed by the industry. In the words of Arne Nielsen, president of Mobil, the undertaking "will mean the drilling of wells which wouldn't otherwise be drilled."

Petro-Can is studying the prospects of participation in foreign ventures in the North Sea, in the Venezuelan oil sands and off the coast of Vietnam. The public is told this is to secure future sources of oil for Canada. However, as long as Petro-Can does not control the marketing and refining of imported crude, Canada is again simply playing into the hands of the oil giants.

Petro-Can is also involved in the Alberta oil sands through the acquisition of the federal government's share of bailing out Syncrude. The bailing out of Syncrude not only represents a potential private loss subsidized, but in terms of the agreement ensures that the bulk of the profits will remain in private hands.

Moreover, Petro-Can has decided to spend $340 million to acquire the assets of Atlantic Richfield Canada Ltd. Under the deal Petro-Can has acquired Arcan's oil and natural gas producing properties in Alberta and British Columbia. These facilities produce about 30,000 barrels of crude oil and 98 million cubic feet of natural gas daily. Despite this production of oil and gas, Arcan has not developed a distribution and refining capacity. In effect, Petro-Can will again be doing just what the private sector wants—developing oil and gas to be sold to the private companies so they can sell it back to Canadian consumers.

The disturbing prospect for the future of Petro-Can is that it will assume most of the risks and expense of searching for and recovering oil and gas while the multinationals will profitably refine, distribute, and market petroleum products. Petro-Can's president, Maurice Strong, has promoted the corporation as a major new source of risk capital. He has called Petro-Can "a catalyst in the petroleum industry." The oil industry will certainly welcome this injection of public funds, provided the terms are right. It is significant that the establishment of Petro-Can and its operating policies to date have not sent Canada's oilmen to the barricades. Nor has there been a vigorous lobby to repeal the Petro-Can legislation.

If Petro-Can is to be more than tokenism, if it is to play a meaningful role, two things must happen. First, Canada must develop a national energy policy which will ensure adequate resources at reasonable prices, developed under Canadian control, for the public good. To achieve this, a course of public ownership and control of the oil industry in Canada must be pursued. Secondly, Petro-Can should be used as an important and significant means to achieve this end and should buy ownership of Imperial Oil, Canada's largest and most profitable oil company, and operate it as a public utility.

By nationalizing Imperial Oil, Canada would stop the flow of dividends out of Canada and end Exxon's influence in Canada. Petro-Can, by owning and operating Imperial, would be a pace setter in terms of energy pricing and development policies. Ownership of Imperial Oil would give Petro-Can access to valuable experience and expertise as well as information which has hitherto been denied our policy makers. It would also thwart the industry's blackmail tactics of an investment strike, that is refusing to develop existing reserves or explore for new ones until they get the price and terms (tax concessions, export rights, leasing arrangements, etc.) they want.

In short, Petro-Can's takeover of Imperial Oil would be a step in developing an energy industry responsible to national goals as set out by the people of Canada through their Government.

Any smaller role for Petro-Can will only perpetuate the mistakes of the past and strengthen the stranglehold the multinational oil companies have on the vital energy resources of Canada. The energy crisis now faced demands public ownership, not government tokenism.

DAVID LEWIS

David Lewis is attached to Carleton University as Visiting Fellow of the Institute of Canadian Studies and as a professor with the Department of of Political Science. This appointment followed a long and outstanding political career beginning in 1936 when he was elected National Secretary of the CCF, a position he held until 1950. In 1962 he was elected NDP Member of Parliament for York South. From 1971 to 1975 David Lewis was National Leader of the New Democratic Party. He is co-author of *Make This Your Canada* (1943) and author of *Louder Voices: The Corporate Welfare Bums* (1972).

7

David Lewis

Canadian Energy Policy: The Collapse of Credibility

Some of the present energy policies of the government were announced by the Prime Minister, if I recall the date correctly, on December 6, 1973. I remember the day vividly. As NDP leader I was in Fort McMurry, visiting the operations of the Great Canadian Oil Sands Company and examining the Athabasca Tar Sands. Knowing that the Prime Minister was due to make a statement on energy on which the fate of the minority government rested, I was on the long distance phone with Tommy Douglas and Stanley Knowles, then as now NDP energy critic and House Leader respectively. I learned that the Prime Minister had apparently studied carefully the statements made by Mr. Douglas and myself on the subject of energy policy and had come forward with a programme that seemed to meet each one of the six points we had been pressing. Among them were demands for a low domestic price coupled with an export tax and a publicly owned petroleum corporation. Tactically we had no alternative but to accept the government at its word, but the doubts we had were soon fully justified.

When I see what is happening to the domestic price of oil, I must admit that to an important extent the NDP was had, for although the domestic price is still lower, the policy has changed: the world price is now the domestic target. And when I look at Petro-Can I am quite sure that this abortion is not what we had in mind. It is this sort of experience—a structure which appears to meet requirements but, in fact, does nothing that really changes the situation—that has worried my colleagues and myself over the years. We had a similar experience with the Canada Development Corporation and I am sure that I could give you a dozen other examples.

It seems clear to me that we must get the Canadian people exercised about the problems with energy policy in Canada. Government agencies and corporations have been getting away with murder for years as far as the needs of Canada, both present and future, are concerned. Unfortunately, it has not been easy to move the Canadian people to see that.

There is an old saying about not being able to fool all of the people all of the time. Well, you are mistaken if you think that it is a universal truth. There is an exception to it. If you are the Canadian Minister of Energy and a member of the National Energy Board in Canada you can and will be fooled all the time. It is part of a long exercise in Ottawa. The oil and gas corporations in Canada have either deliberately lied to Canadians or are totally incompetent. And since they could not possibly have made the billions of dollars in profit by being incompetent, you can reach your own conclusion as to what they are.

It is only six years ago that I heard Joe Greene say, in parliament, supported by a report of the National Energy Board, that we had reserves of oil and gas that could carry us for a century. There was nothing to worry about, he said. Furthermore, in order to make it possible for the corporations to continue living—these poor, impoverished corporations—they would have to export oil to the United States. That's what the minister said. By coincidence it was also precisely what the corporations said. It was exactly at that time that the corporations wanted to be sure they had a market for oil and gas in the United States. They assembled, therefore, figures which showed that Canada had almost limitless reserves of energy. By another incredible coincidence, the Canadian Department of Energy and the National Energy Board came to the same conclusions about the reserves. And these conclusions were similar to those the corporations had used ten years earlier to get the Canadian market split in two at the Ottawa Valley. Joe Greene and his subordinates went to Washington on their knees to persuade the American administration to increase the import quotas on oil from Canada. They did it—the corporations, the Minister, the advisers, all of them—with a great feeling of obligation to Canada, they said. And there were times when they said it with such an appearance of sincerity that I almost believed that they really thought they were doing something good for Canada. It turned out that they were cutting Canada's throat.

Three years later, in 1973, the multinational corporations decided that they had to look for new areas to explore and to exploit and they turned to the Mackenzie Valley. In order to make it understandable, for ordinary people, that the pipeline was needed for gas and eventually for oil, that we really had to tap the north and bring fuels down without delay—even if it meant joining up with the Americans so that the Canadian line would primarily carry gas from Prudhoe Bay in Alaska to central United States

—the corporations came up with figures about the reserves which justified that approach, or so they thought. Consequently, in place of formerly immense reserves in oil and gas, only three years later the reserves were no longer limitless. Reserves were contracting but it was not a serious situation yet. Canada still had enough to go just into the next century. Thus, there was no need to get everybody panicky—yet. Canada could still dream of self-sufficiency. And lo and behold, these new lower figures were confirmed by the Department of Energy and the National Energy Board.

Then the international price of oil shot up. It went up to twelve dollars or more a barrel—instead of the three dollars the corporations were getting in Canada. The corporations decided that they couldn't accept the idea of Canadians getting their own oil at a lower price. It was therefore no longer possible to talk about the reserves lasting into the next century. If that were so, it would make it difficult to argue about exploration, the high cost and the urgent need. The result of this tortuous corporate calculation was that every subsequent year the reserves became more and more contracted.

In 1973 there was enough to last into the next century; in 1976 there can no longer be a policy of self-sufficiency. There can only be a policy of self-reliance. (It truly requires a particular kind of bureaucratic semantics to make this distinction.) And the Minister of Energy once more produced figures that underlined the industry's need to squeeze the Canadian people and the Canadian government into a higher price and new tax concessions.

The people of this country have got to be told again and again that they have been lied to by the petroleum industry and that the government and the National Energy Board have been taken in about the reserves which Canada has. I am no more prepared to believe the corporations and the government in 1976 than I was prepared to believe them in 1970. I am certain we have a serious problem but I am fully convinced that it is not as serious as Mr. Gillespie and the industry now try to make out.

When you get to gas, the National Energy Board said in its latest report that Canada had about 24 years of supply. Even if it is only 20 years, there is sufficient time in which to develop new plans. There is no need to rush into the Mackenzie Valley Pipeline tomorrow. There is time to consider the serious problems. As far as gas is concerned, the government has years during which to further the research work necessary to study the ecological effects; during which to arrange the environment in which the native peoples in the north have to live; during which to settle aboriginal claims. There is certainly nothing to justify irreversible decisions and actions before the Berger Commission tables its report.

There is this strange habit peculiar to the American oil industry which one should observe in passing. Although it doubles its output roughly every ten years . . . it declares every other year that its peak of production has been passed and that its oil fields are well-nigh exhausted.
. . . Nevertheless new pools are continually producing wells brought in to replace those declining, and each year the total output turns out to be surprisingly greater than the year before. One cannot doubt that the lugubrious prophecies of American oil men are in some way related to the wish for higher prices.

The Oil Trusts and Anglo-American Relations, 1923.

There is something else that astounds me. There are three organizations in Canada seeking pipelines: Canadian Arctic Gas Pipeline Limited; Foothills Pipeline Limited, also known as the Maple Leaf Line; and more recently, Polar Gas Limited, which proposes to bring gas down from the eastern Arctic islands. I don't pretend to be an expert in the field and I can't give any technical advice, but I have read a good deal about these outfits and my common sense tells me that if I were the Minister of Energy, I would take a very good look at the Polar Gas proposal.

Why am I saying that? First, because Polar Gas has already found the gas. In the Mackenzie Delta only about four trillion cubic feet have been found. In the islands on which Polar Gas has explored and is working, they estimate that they have already discovered 17 or 18 trillion cubic feet. That is four times as much in gas reserves in the eastern Arctic as has been found in the Mackenzie Delta.

Secondly, there are very few, almost no native communities in the eastern Arctic, compared to the western Arctic. The disruption of the life of the native people would, therefore, be almost nil.

Thirdly, and perhaps most importantly, the pipeline must be totally Canadian. A pipeline in the eastern Arctic would not be anywhere near Alaska so the Americans couldn't play with it. They couldn't bring the gas from Prudhoe Bay and the gas in the eastern Arctic down together because the line would come down the side of Hudson Bay. The line would be Canadian.

The Polar Gas proposal may well not interest Mr. Gillespie who, before he became the salesman of Canada's natural resources, was the Minister of Industry, Trade and Commerce responsible for the Foreign Investment Review Agency, another abortion in government policy. But ever since I read about Polar Gas, I have said to myself, why in heaven's name isn't the government looking at it? Why, when Tommy Douglas suggests in parliament that Polar Gas ought to be given preference, does Mr. Gillespie say "oh no"? I don't know the exact reason. I suspect there may be some overlapping between the companies committed to the Delta consortia and those committed to Polar Gas. I have a suspicious mind where corporate power is concerned because I have spent a life-time of watching them in operation. And I suspect that our authorities will not look seriously at what appears to be a perfectly viable alternative to the Mackenzie Valley Pipeline because they are already talking with Washington. They have been talking with Washington for a considerable time. The government is considering a treaty with Washington, a treaty having to do with safeguards here, there and elsewhere. The ties with the United States over Canadian gas and oil go on continuously and this may well be the reason that the government is not going to look at Polar Gas seriously.

We are told that Canada must increase the price of oil and gas to ensure that further exploration will take place in the very expensive, new areas where exploration is now feasible. I happened to see Alastair Gillespie talking about energy policy on the CBC television programme *This Day* the day after I saw Jim Laxer on the same show. Making my own comparisons, I have decided the wrong man was minister. What I heard Mr. Gillespie say didn't surprise me, but nonetheless made me very angry. When he was asked by the reporter about a guarantee that the increased price would lead to increased exploration, the first answer he gave in support of the increased price was simply—and I am not misrepresenting him—that unless the companies got more money, they just weren't going to explore. I am sure he was telling the truth. I am just heartbroken that he and his government are cowardly enough to be intimidated by the multinational corporations which ought to be made to serve the interest of Canada or be kicked out.

I have said many times since I have become attached to university work that we have a double standard in Canada. Indeed, I used to point this out from the political platform as well. When the poor postal workers go on strike and there is some discomfort for people who don't get their letters and some inconvenience to business, the whole bloody country is up in arms. The high school teachers in Toronto strike. The strike doesn't cause much harm to the students or to anyone else but the whole populations of Toronto and Ontario are up in arms. And any time a group of workers is on strike every person in the media has a horror story. Those strikes, even the serious ones, merely create inconvenience for two, three or four weeks. If the strike is serious enough to cause more than inconvenience, serious enough to interfere with the economy, then government steps in and orders the workers back to work by legislation. Even persons like myself have voted for such legislation because I believed that Canada really needed such action and that the workers weren't going to gain anything from continuing their strike. Yet I believe people should be able to strike without being forced back to work when their strike causes only a temporary inconvenience or hardship. And everybody gets up-tight about that.

But when the corporations that make up Syncrude or other multinational corporations strike, then the government gets down on its knees to meet their demands and nobody in Canada, no commentator, no editorialist, condemns them. When the oil corporations make the profits which they have made and are making, and then refuse to use those profits for the purpose of finding new resources in Canada, they are striking. And they are not merely causing inconvenience; they are threatening the very future of Canada and the welfare of its people. But our present society accepts that. Corporations have a right to do that. They have a right to say we won't produce unless we make x + y profit. It is the worker who has no right to say that he won't produce when he wants x + y wages. This is the double standard under which we live in this society. And since most Canadians are working people, whether they are in unions or outside unions, in the office, or in the factory, or on the land, blue collar, white collar, or professional, they should recognize this double standard, reject and stamp it out.

I looked at past figures to find out whether there is reason to believe that the corporations will use the extra money they will get from the higher-priced fuel to explore for more supplies. Let's look at what they spent on exploration in 1972, before oil prices went up, as compared to what they spent in 1975, as a proportion of their total revenues over those years. The revenue of the oil industry, its total revenue, was roughly two billion dollars in 1972; it was roughly six billion dollars in 1975. In other words, the total revenue was three times as large or up 300 per cent. But the increase in exploration expenditures was only half or 50 per cent; from $500 million to $770 million. And if you think that they may have had other expenses, I looked at another way of checking the exploration expenditure. In 1972 the petroleum corporations spent 23 per cent of their revenue on exploration and in 1975 they spent only 12 per cent or about half as much, proportionately. What guarantee has Mr. Gillespie got that when he raises the price another two dollars that that will induce the companies to spend more on exploration than they have previously spent? None at all.

As a matter of fact, *The Toronto Star* reported that Jack Armstrong, the Chief Executive Officer of Imperial Oil, admitted to the Bryce Commission on Corporate Concentration that in order for Imperial to spend more than five million dollars in Canada they have to get the approval of Exxon, their parent company. Well, if I am to believe the corporations and Mr. Gillespie, five million dollars will do no more than find the place where the hole is to be drilled. Five million dollars in terms of exploration is equivalent to the cost of a cup of coffee in the Chateau Laurier, and that is no longer a nickel. I am not at all surprised at this admission but I find it very serious. If Imperial is allowed to spend no more than five million dol- without permission from Exxon, there simply is no assurance that there will be increased exploration. What influence has Gillespie got over Exxon? I know what power they have over him, but I want to know what power he has over them.

The unfortunate part is that the energy situation in Canada has been an unnecessary and continuous tragedy for the Canadian people. Canadians have been fortunate enough to have oil and gas, large deposits of

I HAVE GOOD NEWS AND BAD, SIR!
WHAT'S THE GOOD?
16/6 ©TSS
THE OIL COMPANIES ARE TURNING TO THE SUN FOR ENERGY
AND THE BAD NEWS?
THEY'RE BUILDING A PIPELINE!
WICKS

coal, and deposits of uranium. Canadians are fortunate enough that from time to time they even have the sun. Canada is thus one of the really blessed countries in this world. There is no reason for energy worries in Canada, none at all, except that the resources—with the single exception of the sun—are in the hands of private multinational corporations whose concern for the welfare of Canada is only an occasional by-product forced upon them by the need to pretend concern in order to continue exploiting the country's resources and its people.

There have been people who have written about the CCF and NDP in the last number of years to the effect that David Lewis is not a doctrinaire socialist, even that he is not a socialist at all. In 1963 or 1964 Walter Gordon, on the other hand, made a speech in the West in which he said that David Lewis *was* a doctrinaire socialist. Since this kind of argument depends on definition, I find it irrelevant. What I want to tell the people of Canada is that ownership and control of our energy resources is not a question of doctrine and never has been, not only with me, but with anybody of common sense who doesn't have a corporate axe to grind. It is plain common sense. No one in Ontario would suggest that hydro electricity generation should go to a private corporation. It was not a socialist government that introduced public ownership of hydro back in 1905. With energy resources, it is exactly the same. Petroleum resources ought to be in the public domain, not for any abstract doctrinaire reasons but because that is the only way in which two essential objectives can be achieved.

Objective one is that energy resources shall remain Canadian. This cannot be guaranteed so long as resources remain privately owned unless there are adequate regulatory laws which are strictly enforced. Secondly, public ownership is the only way in which resources will be developed with an eye to the future of Canada, and the future of the United States and other countries as well. Although Canada can have surpluses, no one is suggesting that we build a wall around ourselves. What I am suggesting is that Canada's future needs as well as its present needs must have priority.

Therefore, I am very much in favour, as I have been for a good many years, of the government taking into public ownership at least one of the large multinationals operating in Canada. I see no reason why Mr. Armstrong should not feel honoured by the proposal that his company, Imperial, be given pride of place on the list. If he doesn't understand the reason, I will explain it to him. It is not only because Imperial is big; it is also because it is an efficient operation. We have had the experience in Canada of taking over bankrupt organizations and making them publicly owned enterprises. It is time to take over a corporation that is really successful.

And since Mr. Armstrong and his colleagues have done such an excellent job, it is right that they should be given the honour of leading the parade to public ownership of petroleum corporations.

I think the Public Petroleum Association is right in advocating the takeover of Imperial. I think it would be too difficult financially and for other reasons to go beyond one major corporation. But if we had one major corporation, we would have a real influence on the petroleum industry.

For the first time we would have direct information so that the corporations would not be able to fool us. For the first time we would be in a position to know directly about costs, about the relationships of the well to the refinery to the wholesaler to the retailer which this huge, vertically integrated corporation presents. No longer would the petroleum industry be free to select its figures on reserves to suit its own corporate interests. By God, it has had it far too easy. The time has come for a fundamental shift of information about and control of the petroleum industry to the people of Canada.

Two Inches Between Me and the Wall

DOCUMENTARY

Small Town Politics: Recreating Feudalism

INTERVIEW BY DAVE GODFREY
WITH DENNIS BARRY, AUGUST, 1976.

Dennis Barry has delivered oil to the farm since we moved there. He is a monthly part of the landscape; when we bought an old oil furnace at an auction and dismantled it and put it back together again in the drive shed so we could work on repairing machinery in the winter, it was Dennis who came up and checked the wiring and pipes and got the controls functioning again.

Erin is a friendly town. Dennis's brother Jeff leads the local band and takes photographs for the local paper. The town has an ethos of success, typified by the story of Stan Leitch, who came to town in the Depression with $37.00 in his pocket and became the Massey-Ferguson dealer and was working towards the last few thousands of his million dollars when he died, from his second stroke, but no one really minds if you don't keep up with Stan.

And nobody questions cheques or asks for a driver's licence; everybody knows who's good for credit and who isn't.

So Dennis was a little apologetic about having to phone me up about an overdue bill.

"Come right up and I'll write out the cheque," I said.

He stood in the kitchen and began to explain. "They just take it out of my commission, you see. They do the billing, but if anybody's slow it comes out of my commission."

He shuffles a little, as though unhappy with the prospect of explaining anything, but when I press him, the whole story pours out. The competence is there. He understands what is happening; it's just that he doesn't see a way out.

"Oh, that's nothing," he says. "That's not the worst of it at all. Every so often I'll get skinned, but it's cash on the barrel for some of them. You get so you just don't care if it's the dead of winter or not. But the bastards have got me tied up tighter than a honey wagon in spring.

"Closing the bulk plant, that was the last straw. I just went in and told them I wanted out."

He says it casually, but you know he has five kids and that this has been his life for seven years and that the agency means a lot.

"I bought it seven years ago, the route, from Harvey Peavoy. Like I bought it out of the earnings, I didn't have any cash to speak of. I just paid him a thousand down and a quarter of a cent each month on the gallonage until it was all paid for: forty one hundred dollars. And Supertest collected it for him; they knew what it was for.

"Well I went into them, that's B.P., when they closed up the bulk plant and said, look, at least I want my four thousand back and they said, 'what for?' What for? said I, for the route, my route. 'You don't own it,' they said. 'We own it.'

When Dennis bought the route, Harvey Peavoy was delivering for Supertest, a Canadian firm that was purchased by B.P. in 1972. There are no records of the purchase available to Dennis, but the commission cheques show clearly that the ¼¢ per gallon was deducted regularly with interest. The value of the route is usually calculated on a quite simple basis: gallonage times 1¢. Harvey Peavoy was delivering 410,000 gallons a year. Dennis is now doing considerably better than that.

But B.P. hasn't chosen to share its increased revenues with him. Dennis Barry receives 3¢ a gallon for delivering gas or diesel fuel. In 1969 when he bought the route he was receiving 2.5¢, the same amount Harvey Peavoy had been getting for at least the previous five years. So that in ten years the rate has increased at an annual rate of only 2 per cent, hardly enough to keep up with inflation, let alone oil company profits. In fact, the jump occurred back in 1972 when B.P. entered into its first arrangement with Supertest. What Dennis had bought in 1969 was a contract running for six years. In 1972 B.P. provided him with a new contract and raised the delivery rate on gas by half a cent and on diesel fuel by a tenth of a cent. And in exchange, although they didn't make it clear to Dennis, they took over any rights he held in the route, even though he was still two months away from paying off the debt.

"But I didn't find that out until 74. You need the supply so you sign what they put in front of you. And they come in with all the new stationary and forms and everything and confuse you. It was in 74 that I found out, when they told me they were closing down the bulk plant here and moving my supplies to Orangeville. So now I had to drive another 14 or 15 miles each delivery. Add that to your expenses and what do you get, eh? And it was right in the Supertest contract that they would keep it open until the contract ended.

"So I told them I wanted out. It just wasn't worth it. And that's when I found out I didn't own the route. So I said, 'well what about the truck, what'll you give me for that?' 'Well,' they said, 'it's depreciated quite a bit.' It was new in 1971, brand new. Supertest used to loan it to you for so much a month. They paid for the licences and you paid repairs. By 74 I'd paid Supertest over $4500 for it. Then B.P. said I had to buy it and pay for the insurance and licence myself; that's another little $800 a year. They gave me $1400 off list then as depreciation, but of course you know they didn't buy the thing at list. They wouldn't let you buy if from a local guy, like somebody you deliver to all the time, no, you had to go all the way to Owen Sound where they got them all and got a deal. So a year later, suddenly the truck wasn't worth anything at all. They owned the tank, you see; they never sold you that.

"'What's it worth now?' I said. 'Maybe three thousand,' they said. 'And what about the tank?' I said. 'Oh, that includes the tank,' they said. 'And it hasn't depreciated much eh,' I said. 'Oh, no,' they said. 'They don't. It's worth at least two thousand.' 'So that's a thousand for the truck,' I said. 'Right,' they said. 'Without the tank.'

"Knowing full goddamn well that there's no way you could get the tank off, with the meters and all, for that kind of money. So the truck was worth nothing. They just keep you tied up. They pay you $240 a month for the truck, on top of commission, once you buy it, but of course it costs a lot more than that, in upkeep and depreciation. I mean look it's almost that for depreciation alone—at their rates.

"Some people just quit. Don Brown from Shelburne did. But here, we're between Guelph and Brampton and Orangeville and Georgetown and they're just waiting to get me out, I think. I went in a while back and he said, 'You've still got your electrician's licence, don't you, Dennis?' I mean what does he know that for. They just want to squeeze you out, like the stations, eh. That's what all this self-serve is about. Squeeze. Squeeze and push.

"I told them I'd just get my stuff from somebody else. Another dealer said he could give me as much from his commission as they are. He's doing over two million. But they said nobody can do that. Nobody gets enough to give me what they do. And you wonder how long he'll keep it, if he is getting it."

We stop for a beer. We're down in his living room now, with the contracts spread out around me. His wife does the book-keeping and has kept all the records. He's probably ahead of most other dealers in that sense. He has a record.

"What will you do, Dennis, if they get tough? I mean if you let us print

Partners in Building a Great City: Business and the Arts

This concert is supported by Imperial Oil Limited.

The Toronto Symphony recognizes and appreciates the sponsorship of this concert by Imperial Oil Limited.

They have shown, in a meaningful way, their belief that Business and the Arts are partners in building a great city.

this article they're not going to be too happy."

He looks about him. He lives in a handsome brick house on Main Street, but the furniture doesn't come from Eaton's College Street. And his last contract was a two year one, from June 74 to June 76. It's now August. He smiles.

"Well, I haven't got that much to lose, eh. I went in and I told them what I wanted. They said they'd give me a little more on the big customers. You see the 3¢ is just for the small ones. They said they'd put the big ones, like Lane's and the County, up to 2.5¢. They only pay 1.75¢ now. I mean less than 2¢ and some of that's a thirty mile trip, each way. But of course they never did anything about it. They always get you to sign, but they just talk and talk.

"I've got friends too, you know. One guy, a trucker, they buy over a hundred thousand a year and he knew what was going on, and he told them, 'The day Dennis leaves, buddy, we leave.'

"But they keep you tied up like a spider crossed with a chiropracter. I mean you have to pay the tax. This one farmer skipped out and left me with eight or nine hundred dollars. And some of that was tax, eh. But I was responsible for that too. He may have claimed the rebate, but I got stuck with it. And I mean B.P. could collect; they've got the lawyers and the clout. They even wanted to control my licence. I have an oil burner licence, a furnace repair licence. I make more out of that than I do out of them, if you figure in expenses. And they wanted me to give that up. So I would only repair their furnaces. Ones people had bought from them, or their customers. I just didn't do that. I couldn't.

"You see, I just don't have that much to lose, there's only so many inches between me and the wall, so what can they do. But they're so cheap, you know. I said, 'why can't you give me more? I mean I used to deliver at 19¢ and now it's 43¢ and going to 47¢.' 'Well, we think it's a fair price now,' they said. 'So, I must have really been making too much before eh, and Peavoy before me, he must have really been making too much.' 'Yes,' they said. 'We have to agree with you there.' The bastards, you know, Supertest used to have a party every year. They knew we worked, every day from October to May, and Christmas Eve until midnight. So they gave us a good party, a motel room and everything, food and drink. But B.P., they had a night out for the agents one night and they gave everybody a beer ticket. One beer ticket and then you were on your own. And Don Brown, you know, the one from Shelburne who quit, they knew he didn't drink so they didn't give him a ticket. Cause they knew he'd give it to me. Now that's cheap. Real cheap."

He laughs. I tell him about some of the figures Bruce Willson and

Walter Gordon and Jim Laxer have provided in the rest of the book. We look at one another. "Goddam," he says. "I'll have to take them up a copy. You know, if somebody would just do an Eagleson. I often think about going to all the dealers and say look, just give me a tenth of a cent on everything extra I get for you. Then we'd shut the whole province down some Christmas morning and people would wake up. But, ah you know, they're all afraid. As soon as you start talking like that they clam up. They're just starting up, or they're getting something under the table, or they've got big plans, or they're selling furnace oil as diesel fuel and pocketing the road tax and they know the company knows and isn't saying anything."

He paces about uneasily. We finish the beer and go back to our jobs. Nothing has been resolved. Nothing made very clear. Change is going to be very hard.

SELF SERVE

3

THE PUBLIC AND THE NATIONAL ENERGY DEBATE

IAN MCDOUGALL

Ian McDougall is a Professor of Law at Osgoode Hall in Toronto. Prior to this appointment he was Professor of Resources Law at Dalhousie University. He is Chairman of the Public Interest Coalition sponsored by the Canadian Wildlife Federation and has appeared at the Mackenzie Valley Pipeline Hearings of the National Energy Board for that body. Ian McDougall is a member of the Legal Committee of the Canadian Arctic Resources Committee, a director-at-large of the Committee for an Independent Canada, and a director of Energy Probe.

8

Ian McDougall

The Mackenzie Valley Pipeline Hearings of the National Energy Board

The National Energy Board attracted a great deal of academic interest of late, and over the past two or three years we have seen some outstanding contributions to the available literature on its history, procedures, composition and effectiveness. I am thinking particularly of the work of Professor Lucas who produced a major study of the Board for the Canada Law Reform Commission, the analysis of Professor John Helliwell which has been the basis of both articles and presentations before the Board, and finally the recent work of Professor Dosman which has lent a disconcertingly full confirmation to allegations of government-industry-Board duplicity.

Rather than discuss the Board's general performance I would like to discuss the current status of the Mackenzie Valley Pipeline Hearings.*

There is an impressive array of "public interest" groups represented at the Hearing, including the Committee for an Independent Canada, the Canadian Wildlife Federation, the Canadian Arctic Resources Committee, the Energy Workshop, the Committee for Liberty and Justice, the Consumers Association of Canada, and the University of Toronto's Energy Probe. There has been, in addition, recognition by the Board of the Native groups such as the Indian Brotherhood, the Inuit Tapirisat, and the Committee for Original Peoples Entitlement. Over and above these groups there is a wide range of other non-industry parties that have standing before the Board, including of course the various provincial governments, and perhaps most significantly, a smattering of major secondary industries and industry-associations. There is, in short,

---

*This paper was delivered to the Public Petroleum Association conference on Nov. 29, 1975.

a wide group of interests that, technically-speaking, has entered an appearance before the Board.

If you are impressed with the extensive array of appearances by parties not allied to the oil and gas industry, don't be reassured for it is only a technical reality, not a practical one. The Province of British Columbia has not showed up thus far. Nor have any of the Native groups. Nor have most of the secondary interests. The public interest concerns named above are represented for the moment by one full and two part-time lawyers. The public interest groups have commissioned no studies and have to date called no professional witnesses. Nor is there any signifacant pool of financial resources which might make a meaningful change to this situation. There are only two cases that are being put forward; that of Foothills Pipe Line Ltd. and that of Canadian Arctic Gas Pipeline.

When we talk about "public interest groups" we may be being slightly presumptious. But of one thing there is no doubt; as Mr. Justice Berger so aptly put it earlier this week, it is in the public interest that responsible publicly-interested groups be heard. In this regard the Berger Inquiry has established its place in history for the priority that it has given to providing the necessary resources to enable effective public interest representation, not as an act of charity or paternalism, but in recognition of the essential importance of having the benefit of more input than simply the well-prepared cases of those with vested interests.

In this context the difference between Mr. Justice Berger's approach to the public interest group question, as against the approach of the National Energy Board is like day and night. The Berger Inquiry is on public record as having fought hard in support of public funding. It is also on record for having successfully demanded notice and disclosure of all government studies and documents pertaining to the pipeline. Its readiness to tailor procedures to conform to the needs of the situation rather than the reverse is becoming legendary.

The NEB on the other hand, far from championing the cause of public interest group funding, has remained conspicuously silent. And it has thus virtually ensured that the public interest cases will be of a low quality. The record at the end of the day can't help but reflect this inequality and the decision in favour of one or other of the lines will be that much easier.

It may be that this result will prove to have been quite deliberate. Certainly some have contended that the inaction of the Board and the refusal by Mr. Gillespie to provide resources to public interest groups to prepare their cases is aimed primarily at ensuring an easy decision. In other words, the Board wants to be in a position where it can point to the record and say that the public interest groups were more of a nuisance than a help and that the case in favour of a line was largely unopposed. I personally cannnot testify as to the merit of this logic, but I can certainly assert that there is no evidence at all to oppose it.

This Hearing involves a project representing the largest single commitment of private capital in the entire history of the country. It has been compared in importance with the CPR. This is a good analogy because before the project is over the taxpayers of Canada will have to hand over a king's ransom to carry it to completion. The importance of the NEB Mackenzie Valley Piepline Hearing simply cannot be overstated. So much depends upon it. It is Canada's last clear chance to regain control over its energy fate. This is a crucial point and there is every reason to believe that government planners have failed to give it adequate consideration.

Studying and appearing before the Board has for a number of public interest intervenors been rather a shaking experience. In my own case it all came about quite innocently. I was given money by Shell, Imperial, the Canada Council, and one of the American law schools to write a thesis on the Board which was completed in 1971 and submitted in 1972. The Board was of interest to me, as it was to others, in a quite technical sense—namely whether it was effective in fulfilling its mandate and what lessons, if any, would a review of its track record have for future reform. I had no expectations of discovering that the NEB would prove on scrutiny to be the monumental failure that it in fact is. There was no reason to think in advance that the most reasonable judgment after analysis would consist of two words: "rip off". Indeed it is doubtful that Shell, Imperial, or an American university would have funded a study that would conclude by arguing for a termination of U.S. exports.

For the handful of non-political scientists who have examined the NEB it has been a disheartening, and at times mind-boggling education, for the basic lessons have been broadly political rather than neatly technical. What I mean to say is this: working out a better export formula isn't good reform; changing the composition of the Board, while necessary, isn't sufficient; and buying a later generation computer for market research won't help. None of these piecemeal things will do. What is needed is a tribunal that can hold a hearing without walls and produce a report without varnish. For this, the most basic changes to the very underpinning philosophy of federal regulatory tribunals is required.

The walls of the hearing are not simple plaster and mortar. They are the much more substantial fiscal, formal, structural, and technical barriers that keep out all of those persons that have the most to say, and ultimately the largest stakes at issue, that is the public itself. Few of us can leave our jobs without pay for 18 months. Few of us can raise upwards of a million dollars to produce studies, lawyers, and witnesses. Few of us are well-versed in the rites of the legal profession. Probably only a handful of us know the principal actors in the Board or in the industry. And how many of us are at ease with the basic technical aspects in controversy such as Monte Carlo Summation, Residential and Commercial Demand Elasticity, Cross Elasticity, Deliverability Profile Assessment, Line Costing, Recovery Control, Minimum and Maximum Rate of Take Provisions, and so on? For, make no mistake, this nonsense is the common denominator of a Board Hearing, not the general gut issues of the public interest.

When I say that we must have unvarnished Board reports, I mean that they must exhibit candor, admit uncertainty, expose risks, and be critical in language that the public can readily understand and, if appropriate, act upon. For too long the Board has masked its movements and obscured its mistakes behind undecipherable jargon, and prevented the recognition of a basically straightforward crisis by the citizens of Canada. Everyone knows how to react to a future that promises $2.00 gasoline, $4.00 natural gas, and rations upon home heating, driving and flying. But not all of us are sure how to respond to burner tip parity, commodity value, reserve blow-downs, and deliverablility shortfalls. You've got to be dumber than a bagfull of hammers not to understand what a government body means by "we haven't taken the time to examine what might happen to taxes, income, trade, employment or welfare if our recommendations are adopted." However, an expression such as "at present the selection of a demand elasticity forecast between the case A, B, and C estimates is problematic and must await further consideration" is hardly what one would call panic inspiring, or even meriting a brief note to the local MP.

The NEB's primary responsibility is to us, the public. The public pays the NEB's salaries. The NEB have been playing fast and loose with Canadian oil and gas, and ultimately our well-being. The Board is accountable to us for the fact that it has consistently made a near gratuitous donation of our birth-right to the multinationals and the export markets that they serve. When did Canadians last have a plebiscite on the export of its cheapest southern gas to California? What we did instead was to irrevocably bind the Board to the public interest through the sovereign force of statutory enactment. An it's now obvious that both changes and an explanation are owing without delay.

And when I say change is needed, I don't think in terms of a leisurely, thorough perusal of the National Energy Board Act, the background of the Board members or the efficiency of its staff. The Board is only important if there is petroleum and gas to regulate. The first step

strong · emotional

therefore must be to ensure that Canada's only too limited conventional supplies are preserved.

The truly amazing thing is that there should be a debate about oil and gas at all. How much does it take to convince people? Consider the hydropower, uranium, and coal, in addition to the oil and gas, developed prior to the Board's establishment in 1959.

American consumers have absorbed much of the cheapest oil and gas that Canada will ever be capable of producing. The United States is also deriving ever-increasing benefits from large Canadian hydro-electric developments and Canadian storage projects. The massive dams on the Columbia, Kootenay, Nelson, Churchill and James Bay river systems all illustrate Canadian energy development in a North American context. So, too, do the major gas and oil pipelines, including Trans Mountain, Interprovincial, Westcoast Transmission, Alberta and Southern, Canadian Montana, and TransCanada.

Every major power and fuel reserve within commercial reach of the 49th parallel is either already exploited, or is in a state of mid-development. The point would seem to be only too clear. The "continental energy package" is a full-blown reality, not a hypothetical abstract. All the necessary infrastructure—from the producing machinery to the transmission lines—is already in place. What remains is a nation of much reduced potential in terms of the future development of cheap energy sources, and all of the economic and welfare benefits that derive therefrom.

The Canadian energy crisis is a crisis of our own making. We would not at present have a crisis but for the steadfast refusal on the part of the federal government to apply effective controls over the rate at which we produce oil and gas from only too limited proven reserves. For too long we uncritically accepted the proposition that there were boundless supplies of oil and gas just waiting for development. For too long we also accepted the proposition that private industry must be encouraged so that it might help us realize the benefits accruing from the hypothetical treasure house. And thus for too long private industry was allowed to produce and export from the solitary source of petroleum and gas in Western Canada.

In 1974, 48% of oil production and 42% of natural gas production was exported. If this continues, Canadian natural gas consumers will experience shortages in 1976. In addition, Canada will shortly start running out of petroleum and reliance upon vastly more expensive imports will move progressively further to the west of the present Ottawa Valley line. Given such circumstances, it will obviously be of critical importance to make an immediate start upon projects—such as the Mackenzie gas

Milestone Proclamations in Canada's Continental Energy Strategy

In 1965 the Merchant-Heeney Report on Canadian-American relations commissioned by Lester Pearson and Lyndon Johnson referred to "the economic advantages to both countries of disregarding the boundary for energy purposes, that is, in the development and distribution of energy on a regional north-south basis where this is to the mutual advantage." The report concluded that "we believe there would be virtue in having a joint look at the energy picture as a whole."

In 1969 at a press conference in Washington, Energy Minister J.J. Greene declared himself in favour of a continental energy resources deal so that "people will benefit and both countries will benefit, irrespective of where the imaginary boundary goes."

In 1970 the U.S. Cabinet Task Force on Oil Import Quotas, the Shultz Report, called for a "harmonized energy policy" with Canada. The justification for this was the security value of Canadian supplies.

The report stated: "The risk of instability or animosity is generally conceded to be very low in Canada. The risk of physical interruption or diversion of Canadian oil to other export markets in an emergency is also minimal for those deliveries made by inland transport."

In 1972, Energy Minister Donald Macdonald wrote a letter to U.S. Interior Secretary, Rogers Morton, which highlighted the security advantages of a possible Canadian route for Alaskan oil. Macdonald's letter stated:

"There would be many advantages arising from the use of a Canadian pipeline route. We believe it would enhance the energy security of your country by providing an overland route for your Alaska oil production, thereby servicing the oil deficit areas of the mid-continent and also the Pacific North West.

"Canada has an interest in the energy security of your country, and this land route for Alaska crude oil would enhance that security of supply to deficit areas in the United States. Furthermore, this security of supply could be further enhanced during the interim period of northern pipeline construction by extra Canadian crude."

In 1973, W.O. Twaits, then Chief Executive of Imperial Oil declared:

"It is as foolish to suggest that these two countries (Canada and the U.S.) can follow unrelated or uncoordinated energy policies as it is to suggest that the countries of western Europe, which are interlinked in similar ways, can proceed separately in matters of energy."

In 1973, Prime Minister Trudeau told Canadians:

"Far from reducing oil exports to the United States, Canada has been shipping to the U.S. more oil than ever before. Our projections for this year indicate we will ship to the U.S. 63 million barrels more oil than they purchased from us in 1972—an increase of 18 per cent. By increasing Canadian production, we have been able to respond to our neighbour's needs—while at the same time meeting our first responsibility, which is to ensure basic fuel requirements for Canadians. Consistent with that responsibility, the government's policy is—and will continue to be—to supply the U.S. from domestic oil production with all the help we possibly can. We are friends, and friends care about each other's problems.

". . . rapid changes of the seventies now demand new policies. The most important of these changes includes doubts about availability of oil from foreign suppliers, spectacular price increases for that oil, and soaring new energy demands by our chief oil customer, the U.S. This combination—of less security, higher prices and greater demand—has led to exciting new development possibilities for relatively costly and remote energy sources—such as those in the huge Alberta oil sands, the Arctic and the Atlantic Continental Shelf."

pipeline, more tar sands plants, new hydro-electric developments, nuclear and coal thermal generators—all over the country.

If Canada were prepared to phase out some of its exports on the other hand, the panic would be effectively decapitated. Take natural gas. Here the elimination of but one of the four export systems—Alberta and Southern which sells Canadian gas to its United States parent, Pacific Gas Transmission, which in turn sells to its parent, California's Pacific Gas & Electric—would create enough reserve gas to satisfy existing requirements east of Alberta to about the year 2000.

The proposition that I want to put forth is clear. The exports level is not only more than Canada can afford; but their level is the sole means of buying time. If we don't want to build Arctic pipelines now, if we don't want endless proliferation of costly and dirty energy projects when the only alternatives are both expensive and environmentally destructive, and if we don't want to place the country and its people under nearly impossible economic burdens, then exports must be reduced.

There is an expression that very much applies to our present circumstances: "those who refuse to learn from history will be doomed to repeat it." As suggested earlier, Canada has a rich history of unparalleled resource management blunders to draw upon. For example there is the Columbia River Treaty which is a waste of one of the largest resources of economic potential in a basin capable of generating 15% of the world's electric power.

When one looks at these developments, it is a small wonder that the United States would be surprised if Canadians reacted negatively to water export proposals and continental energy plans. While many object to the precepts of "resource continentalism", we need only look about us, from one end of the country to the other. What we see are projects like the Columbia, Churchill-Nelson and James Bay. What do we think these projects really are? Thus it becomes clear that the present oil and gas export controversy is the latest addition to a long list of dubious energy trade deals. If the debate over the Mackenzie Valley Pipeline seems vaguely familiar it is with good reason.

I am sure there are a few who would say that talk of shortage is scare talk, that there are alternatives such as the tar sands, the Mackenzie Delta, the Eastern Arctic and offshore areas, that gas pipelines don't pose the problems involved with hot oil lines, and that there is great potential for new technology.

There are a number of things that should be pointed out about such arguments. The tar sands have enormous potential, but extracting significant quantities of oil from this source is a task equivalent to drawing blood from a stone. The sheer magnitude of the capital costs of synthetic crude plants guarantee the most expensive petroleum in history. The interest component alone will approach $4.50 per barrel. When refining, transportation, operating costs, and profit are added, we are talking about tar sands petroleum at $25.00 a barrel. What's in short supply is not oil, but the money necessary to obtain the oil from the tar sands. Even if the funds were available, 2.5 billion dollar plants are not put up and producing overnight. Years of lead time is required. Finally, the extraction process is an environmental nightmare. Water must be applied to filter two tons of bituminous sand for each barrel produced. In other words, ten plants would involve the filtration of 2,500,000 tons of overburden each day.

The same situation is true of the hypothetically vast Arctic and offshore gas reserves. The Mackenzie Valley gas line will cost not less than 7 billion dollars. Interest charges alone on each unit delivered will be two times higher than the price at which British Columbia natural gas was sold to the United States a few short years ago. The total cost will be not less than twenty times present Alberta lifting costs.

As well, the potential of the frontier is purely hypothetical. There is growing doubt about the true magnitude of these reserves. It now appears that visions of a vast northern treasure house are just that—visions. After a decade of effort, little has been found in the Mackenzie Delta. Likewise the Eastern Arctic has been disappointing. Many of the East Coast offshore drilling programmes have all but ended due to consistent failure. The B.C. offshore has been written off.

The industry is fond of characterizing its work as "boom or bust". The exploration situation appears to be more or less weighted on the "bust" side of the equation. But while the tally of Arctic finds may be small, if present policies are permitted to continue, Canada will have no choice but to make the heavy investment necessary to get these supplies to Canadian consumers. Without a dramatic change in planning, the Mackenzie Valley Pipeline (in particular) will have to be built on a crash basis. There will be no time to consider the outstanding environmental issues, no time to negotiate a Native Lands Settlement, and no time to debate and consider ways of preventing a rapid displacement of the traditional economic conditions north of the sixtieth parallel.

We are once again on the threshhold of committing the country to what is perhaps the most risky, expensive, and at this moment, thoroughly unnecessary project in Canada's entire history. How is it possible that a Mackenzie Pipeline is in the national interest when in 1974 42% of conventional gas production was exported to the United States? Why is

Does Canada Require the Pipeline Now?

We believe that the supply situation from existing southern reserves is much more favourable than was realized a year ago when the NEB made its gas supply report and Canadian Arctic Gas presented their "crisis of supply" scenario. What evidence do we have of increased supply?

1 New discoveries in the foothills region (such as Shell's Rosevear and Willson Creek).

2 Increased prices have increased the recoverable reserves in existing reservoirs and made smaller pools economic to develop.

3 The development of the shallow gas Suffield block by the Alberta government with proven reserves of about 2 Tcf and growing.

4 Hidden reserves under contract but not being tapped. For instance, Alberta and Southern (Pacific Gas and Electric of California) have 9.7 Tcf under contract but export approval for only 5.1 Tcf. The remaining 4.6 Tcf could be made available to Canadian customers and is larger than the total reserves (proven and deliverable) in the Mackenzie Delta.

5 The Alberta Energy Resources Conservation Board, in its mid-1975 report, announced that larger than expected net additions to supply had been achieved. Expected additions — 1.7 Tcf; actual — 5.2 Tcf; production 2 Tcf; net increase in reserves for 1974, 3.2 Tcf.

6 Premier Lougheed's announced intention to consider releasing some of the existing reserves committed for Alberta use, in return for guaranteed delivery at a later date of frontier gas to Alberta.

7 The federal government's announced intention to cut back on exports of natural gas to the United States (currently running at about 1 Tcf per year or 40% of production).

None of these individually provide the salvation for gas supply; however, collectively they show an additional supply adequate to cover projected shortfalls (about 10 Tcf) through to the late 1980's. We will not need the Canadian Arctic Gas pipeline to heat Canadian homes in 1981.

SOURCE: "The Mackenzie Valley Pipeline and Canadian Priorities," a presentation by Dr. Robert Page on behalf of the Committee for an Independent Canada to the House of Commons Standing Committee on Natural Resources and Public Works, February 5, 1976.

it that cheap Canadian gas should be sold to markets like California when its sale means that Canada must assume all of the economic and enviromental burdens associated with having to develop distant and expensive supplies?

The National Energy Board was set up in 1959 to prevent the very situation we now confront. Gas export was to take place where there was a surplus in relation to Canadian needs. Moreover, the Board was given power, under section 17 of the National Energy Board Act, to vary an export whenever it appeared to the regulators that the domestic interest might be adversely affected. The time has come for this power to be exercised. I submit that selective elimination of exports can prevent the necessity of an Arctic pipeline for many years.

This suggestion is not offered without some appreciation of the international complexities involved in future Canadian-American relations. At the same time the problem should not be overstated. The media has carried ominous reports of possible U.S. retaliation if Canadian oil or gas exports were tampered with—namely cutting off the U.S. section of the oil pipeline to Sarnia, and placing an embargo upon the supply of U.S. coal to Ontario Hydro. Both are absurd propositions. The southern segment of the oil line to Ontario is also a major source for U.S. customers along its route. Stopping the flow would be the classic case of "cutting off your nose to spite your face" because the U.S. would lose three barrels for Canada's one.

As for Appalachian coal, it is imported into Canada for power generation through a system which exports finished electricity into the U.S. Likewise British Columbia, Manitoba, Quebec and New Brunswick are power exporters. To disrupt coal supplies we thus jeopardize an enormous source of power for American use.

I do not mean to pretend that the choices are easy. Rapid termination of exports to the U.S. is a virtual guarantee of the most protracted federal-provincial-diplomatic—and private industry dispute in our history. Difficult as this may be, two factors are clear as crystal. First, even with the recently-announced border price increases, Canada is exporting its one sure source of supply at bargain-basement rates. All Canadians get in return are expensive and hypothetical supplies at best. Second, when all is said and done, these resources belong to Canadians.

We must ask ourselves, what does independent sovereignty amount to if the public cannot marshall the country's internal governing forces to manage resources in the Canadian interest?

ROBERT PAGE

Robert Page is an Associate Professor of History at Trent University. Over the past two years he has been a leading intervenor and witness for the public interest at public hearings on energy including the Natural Gas Supply Hearings (1975), Ontario Energy Board (1975), the Berger Commission (1974), and the Mackenzie Valley Pipeline Hearings at the National Energy Board (1975-76). Chairman of the Mackenzie Valley Pipeline Task Force of the Committee for an Independent Canada and former Chairman of the CIC (1974-75), Robert Page is also host and producer of the public affairs TV series *The Editorial Page*, vice-president of the Peterborough Progressive Conservative Association and co-editor of *Canada Since Confederation: Essays and Interpretations* (1972). He is on the National Executive of the Public Petroleum Association.

9

Robert Page

The Public Interest and Energy Decision-Making at the National Energy Board

In considering the evolution of energy policy in Canada it is essential to separate the three different levels of decision-making. The first level, the local project, can vary in size from a single oil well to a large project like Syncrude; the second level is the wider corporate strategy and planning of the private sector for exploration, production, pricing, etc.; and the third level is the role of the provincial and federal governments in creating policy and carrying through the regulatory process.

Canadians can get so tied up with analysis of specific projects, such as Syncrude or James Bay, that they lose sight of the structures which are more crucial than the individual projects. If the Mackenzie Valley Pipeline was stopped Imperial Oil would dream up new schemes to widen their corporate empire. Imperial by name; imperial by nature. It is equally important to remember that a change from private to public ownership without an equally fundamental change in government structures might leave us with nearly as many problems. As the James Bay experience has shown, the tyranny of government bureaucracies can be as ruthless as that of the multinational corporations.

In this paper I propose to consider principally the third level mentioned above. My experience in the last two years has been in working before government regulatory boards as a spokesman for the "public interest" sector. When one understands the nature of regulatory boards such as the National Energy Board it is easier to see why Canada has made mistakes in energy policy in the past.

The idea of a national energy board had its genesis during the last days of the St. Laurent government when the Gordon Royal Commission recommended a new national authority with advisory and regulatory powers. Early in 1957 John Diefenbaker, as Leader of the Opposition, supported the proposal and expressed his own fears about the extent of American investment "uncontrolled for the benefit of Canada." He believed that changes were needed "whereby Canada in the days ahead will remain an independent Canada and will not inexorably drift into economic continentalism."[1] Shortly after gaining power the Conservative administration appointed the Borden Royal Commission on Energy and one of its recommendations led to the formation of the NEB. In operation, however, this new body did little to carry through the earlier rhetoric of Diefenbaker.

The NEB is currently composed of seven members, any three of whom compose a panel for a particular hearing. They are appointed for seven year terms, must be Canadian citizens, and must *not* be shareholders, bondholders, directors or officers of any company in the energy field. The Board is a court of record with exclusive jurisdiction over matters covered by the NEB Act and possesses all of the powers of commissioners under Part I of the Inquiries Act.[2] Its principal concerns relate to interprovincial supply and pricing, export licences, and certificates of "Public Convenience and Necessity" for energy transportation systems such as pipelines. The NEB may act on its own initiative or upon the request or application of others. Its powers, if it chooses to use them, are quite wide. It may recommend any measures within federal jurisdiction which "it considers necessary or advisable in the public interest for the control, supervision, conservation, use, marketing, and development of energy and sources of energy."[3]

Before becoming law all decisions and recommendations of the Board are subject to Cabinet approval; except that any project rejected by the NEB cannot be reconsidered at Cabinet. Unless specifically raised, there is no discussion or approval of Board decisions in parliament. Its decisions may be enforced as decrees of the Exchequer Court or any provincial superior court. Decisions can be appealed through the federal court to the Supreme Court of Canada on questions of law or jurisdiction.

The rules of procedure vary from hearing to hearing but are generally more relaxed that a court of law. Any party wishing to receive intervenor status must file in advance a statement outlining the basis of their interest in the case. Groups do not have to be represented by legal counsel to speak but most choose to do so.

In the early days the primary role of the NEB was to secure the orderly marketing of the excess production of oil and natural gas in interprovincial and international trade. It developed very quickly a cooperative philosophy with industry and, because it did not force disclosure, was dependent upon the oil companies for the essential baseline data from which it derived its decisions. Data in areas such as reserves was vague and incomplete. Canada was trying to sell as much oil and natural gas in the United States as Washington would accept and Canada west of the Ottawa River was reserved as a protected market for Alberta irrespective of the low world price for offshore supplies.

Suddenly, in the early 1970's this stable picture changed drastically when the OPEC cartel announced its price increases. As world prices shot up the majors in Canada demanded a similar increase in Canadian prices. OPEC created the greatest bonanza for the Canadian producers since the Leduc discoveries in 1947. Up to this point they had argued for protection from low world prices; now they reversed their stand saying that the "free market" forces must determine Canadian prices. Petroleum now became a very expensive commodity where price bore no relation to the costs of production.

These events created new roles and problems for the NEB. With supply no longer assured it now became necessary to restrict exports. The NEB was expected to develop a clear and comprehensive national energy policy with little coherent direction from the government; it was poorly equipped, with limited staff and a laissez-faire philosophy, to accomplish this new task. As governments and the corporations shared the massive economic rents which accrued from the price increases, the Board suffered from a basic conflict of interest in terms of protecting the consumer concern for low prices.

The nature of the NEB is most clearly revealed through its failure to protect Canadian petroleum supply in considering export contracts in the late 1960's and early seventies. The pressure from the corporations for exports was understandable for this was their means of maximizing immediate profits; they presented inflated reserve estimates which misled the Board and the Cabinet.[4] They were able to accomplish this through the confusion over the three categories of reserves which was a product of the lack of full disclosure. The three categories of reserves were "proven", "probable", and "possible". All parties did not have identical definitions, which added to the confusion. Generally, proven reserves were relatively "hard" estimates derived from drilling results, while probable and possible reserves were surface estimates based upon geology and seismic tests. This "volumetric" method estimated the number of cubic miles of sedimentary rock in an area; in theory each cubic mile should contain 46,000 barrels of oil and 276 million cubic feet of gas.[5] It was on this method that the massive reserve figures of 1970 were based; a mathematical equation which has proved to be dreadfully wrong.*

* Companies such as Dome Petroleum still use such techniques in presenting reserve figures to the Board for areas such as the Beaufort Sea.

Oil Reserves: They're all in

Imperial Oil Ltd.

W.O. TWAITS,
Imperial Oil Ltd., 1972:

In the current debate, the export of Canada's energy resources is being questioned; in effect, we are being urged to 'bank' our petroleum resources. Canada is not in any way deficient in energy resources. Our present reserves, using present technology, are sufficient for our requirements for several hundred years.

the eye of the beholder...

Imperial Oil Ltd.

J.G. LIVINGSTONE,

Imperial Oil Ltd., 1976:

. . . Certain predictions are not intended as definitive statements. Their purpose is to outline possibilities. You might say their time never elapses because they are subject to constant updating in the light of new knowledge and experience.

Assessments of undiscovered petroleum reserves are precisely this kind of prediction. They simply describe the hydrocarbon potential of a given area without saying how soon, if ever, the potential will be realized. They serve primarily as a guide to the explorer by indicating how much oil and gas *might* be discovered after all exploration has ended, a period that can extend many, many years. Assessments are limited to whatever data are available at the time they are made. They are imperfect by definition, and as time passes are revised to reflect new information acquired through exploration and development.

. . . Unfortunately, there has been a tendency in Canada to misunderstand the true nature and purpose of petroleum assessments. They have been looked upon, mistakenly, as statements of the reserves of oil and gas that could be discovered and developed whenever required. The industry has now become painfully aware that its past assessments, as well as the assessments made by other agencies, have been misinterpreted by many in this manner.

Yet the Minister of Energy and the NEB seemed to have little awareness of how little these figures meant. It was hardly surprising that such figures were put forward by the Canadian Petroleum Association. It was, however, astounding that even Joe Greene, the Minister of Energy, accepted them at face value. As advisor to the minister, the NEB played a central role in this false scenario.

The NEB does not act alone in advising the government. Senior Board staff serve on key government committees where they influence and in turn are influenced by consideration of government policy. From 1968 through 1970 the Board representatives participated in crucial deliberations of the interdepartmental Task Force on Northern Oil Development which decided that the Mackenzie Valley Pipeline should be built.[6] In addition they forced a merger of the two competing study consortia to try to avoid having to choose between competing applicants.[7] This type of *in camera* activity is seldom known or discussed in parliament, although in this case ministers talked in public of their endorsement of a Mackenzie Pipeline. Acting Prime Minister Mitchell Sharp stated in the House:

> It would be our wish that a gas pipeline be built; I think it would be in the interest of Canadians as well as Americans.[8]

In the light of these committee actions and government statements, is it possible for the National Energy Board to question today whether the Mackenzie Valley Pipeline should be built? It would appear from the record it had made up its mind five years before it began to hear evidence.

The NEB, in considering the Mackenzie Valley Pipeline project, has confined its scope to judging only which of the two applicants presents the best means of delivering Delta gas to southern markets. At this stage there does not appear to be any agency considering the more basic question of whether we need frontier gas now; and if so, which of the frontier areas could best supply that gas.

It is very clear why Imperial and the other Delta producers have opted for the Arctic Gas project: it is because it ties in the Alaskan reserves of the parent Exxon Corporation. It is obvious from the size of the Delta reserves and the routing through Canada that the Arctic Gas proposal was primarily designed as an American delivery system to get Alaskan gas to the lower 48 States. Hence, Imperial is determined to ensure that there will not be serious consideration of the Polar Gas project (which is backed up by much larger Canadian reserves in the high Arctic islands) until Arctic Gas is under way. TransCanada PipeLines, a partner in Arctic Gas, is project manager for the Polar Gas project and has ensured that the Polar Gas application stays in line behind the Arctic Gas application. This "voluntary" approach to energy planning is an abdication of national sovereignty which results from foreign ownership of our key energy reserves, and an ineffective regulatory process dependent on corporate applicants and incapable of independent initiatives.

One of the staggering problems of the current energy scene is the massive capital requirements of all the projects now on the drawing boards. The factor of capital alone will force the government to be selective and to be more interventionist in the energy field. However, today the NEB does not consider the comparative advantage of alternate capital usage. Gassification of southern coal or a new tar sands plant might be a more profitable use of capital. Canada might even be further ahead to spend capital in secondary industry and accept an increased dependency on OPEC. Each proposal must be considered in terms of the economic rents it produces for Canada. A Mackenzie Valley Pipeline costing twelve to fifteen billion dollars, creating 400 permanent jobs, and adding only a small additional gas supply for Canada should obviously have a very low priority. The economic rents of the Mackenzie Valley Pipeline will go to the foreign-owned producers while Canada will suffer the dislocations of financing the project. Self-sufficiency arguments are currently being used by the producers in an unscrupulous way irrespective of the capital requirements of Canada in other sectors of the economy.

In agencies like the National Energy Board there is considerable confusion about the meaning of the words "public interest". Many corporate spokesmen have argued they represent the public interest as they provide jobs and the products for society: what is good for General Motors is good for Canada! Politicians argue that they represent the public interest because they are elected, but on energy price increases governments have a conflict of interest as they cream off huge sums of additional revenue. The emerging public interest groups view their position as a type of class action in confronting big business and big government in the interests of the individual citizen and consumer. Governments have chosen to make public policy through the adversary process of regulatory boards. Therefore, the public must be represented as a party to those hearings to combat the assumptions of private and public bureaucracies. Because the corporate applicants alone can properly document and argue their case, boards tend to become very sympathetic to the industries that they are designed to regulate. Unless Canadians are prepared to finance and support the public interest intervenors, this problem will continue to escalate.

Although five years ago Pollution Probe and the Ontario Waffle were

A Modest Tax Proposal

Canadian society has been much criticized of late for the slothful and unbusinesslike character of its people. Many after dinner speakers have pointed out the need for a return to the work ethic and have warned against a slide into a welfare state torpor.

All this philosophizing has merely located the problem. What is needed is a cure. A simple change in the tax structure may well be the answer.

Imperial Oil has always symbolized businesslike performance in Canada. If every Canadian family could feel that it is a tiny scale model of Imperial Oil, the needed change might be brought about.

Why not put every Canadian family on a tax footing identical to that of Imperial Oil? Given the same incentives an oil company has to improve its performance, wouldn't Canadian families take the bit in their teeth with equal fervour?

The essence of the proposal is this: recognize the living expenses of wage and salary earning Canadians as business expenses.

Under the reformed system, the following features would be in place:

1 The cost of food, gasoline, residential upkeep, education, and health care would be written off as business expenses—in exactly the same way Imperial Oil writes off analagous expenses on its corporate income tax.

2 A capital cost allowance would assist family corporations in the purchase of houses, cars and summer residences. Imperial Oil enjoys such a tax provision for the renewal of its capital equipment.

3 A depreciation allowance would allow families to write off the cost of their houses, cars and summer residences over a 10 year period at 10 per cent a year. This is a normal business procedure.

4 For Canadians unfortunate enough to be out of work a new tax provision would replace unemployment insurance and all the agony that UIC has caused. Such Canadians would be categorized as "engaged in exploration" activity—exploration for employment. They would be entitled to write off 100 per cent of their exploration expenses and would earn an additional thirty-three per cent write off as an earned depletion—exactly the same provision Imperial Oil enjoys.

5 Against income taxes, a twenty-five per cent write off would be initiated to cover the costs of raising children. This is analagous, of course, to the resource allowance now enjoyed by Imperial Oil.

Once these measures were in place, the results would be much the same as they have been with the oil industry. The average Canadian corporate family would pay no tax at all.

The system has worked well for Imperial Oil. Why not give Canadian families the incentive to get on with the job?

threatened with exclusion, recently all public interest groups have been admitted although not taken seriously by Board members; they are tolerated on good behaviour. Recently, Board Counsel for Ontario Energy Board felt it necessary to explain to an American witness that although the Committee for an Independent Canada was party to the Hearings he had nothing to fear because their lawyer was really quite a civilized chap.[9]

Thus far, the public interest groups have tried to get their points across mainly through cross-examination of corporate witnesses rather than by leading evidence and presenting panels of experts of their own. The latter is usually exceedingly expensive and beyond their resources. The costs of participating in these Hearings are extremely high and any analysis of the procedures shows that they are clearly designed to meet the requirements of the corporate applicants, not the public interest intervenors.

The NEB tries to exclude what it terms "politics" from its proceedings but has an obvious double standard. The right wing ideological assumptions of the "free market" and the private enterprise economic system are continually presented in the rhetoric of the industry and even in the econometric studies. If any of the public interest groups should challenge these assumptions they would be ruled out of order as being "political". In fact there is a curious ambivalence in the Board's own philosophy. It seems to be embarrassed to carry through with enthusiasm its own interventionist role for fear of casting doubt on the viability of the free enterprise system.

In Canada in the last few decades we have passed more and more crucial decision-making from the House of Commons to regulatory boards. Although parliament may have difficulty with highly technical matters it does possess the checks and balances of the opposition forces. When the boards were established no such democratic checks and balances were built into the structures. A system has been designed which is an adversary process but there is no provision for one side of the case to be presented. This vacuum is a major weakness in the democratic process in Canada.

In Canada today millions of dollars of public funds are spent annually on government research on policy questions; but seldom does this research get to regulatory boards such as the NEB. In theory the Department of the Environment cannot testify before the NEB because this would constitute one department intervening against another (i.e. the Queen vs. the Queen). As a result this research is largely wasted unless the public interest groups learn of its existence and subpoena the documents. On the pipeline, the government departments have been far more secretive about their plans and research than Canadian Arctic Gas or Foothills. Requests to the Hon. Judd Buchanan, the Minister of Indian and Northern Affairs, for release of such materials were rejected even though the same reports were released to the U.S. Department of the Interior. Similar requests made through the Board have been rejected. The result has been that the public interest groups have had to waste precious funds to duplicate work already done by the government with public funds. How does it serve the public interest that these materials are not presented to the Board as part of the filed evidence to be considered?

This research problem is compounded in Canada by the way in which the private research institutes—such as the C.D. Howe Research Institute and the Conference Board—are locked into the multinational corporate structure and not free for "arms length" independent research in a number of areas. When we in the Committee for an Independent Canada approached the largely government funded Institute for Research on Public Policy we were rebuffed. Their concern to attract key figures from the corporate sector to their board, such as William Wilder of Arctic Gas, and their slowness in getting down to any serious research is certainly discouraging and it would appear that they have no plans for public interest involvement.

Until the public interest sector can obtain financing, these problems will remain. The Northern Assessment Group (the Berger Commission) which drew evidence from Canadian Arctic Resources Committee and other public interest groups at its hearings in the Northwest Territories showed the positive results when serious funding is available. Currently, academics and others are expected to do research on the basis of their own resources of time and money. As well, these citizens organizations are the only "pressure group" whose expenses are not tax deductible. It is time for a change in the system. Public interest research should be funded as part of the regulatory process. One possible approach would have each corporate applicant pay a public interest tax equivalent of two per cent of the total cost of the corporate planning involved in the project. For instance the ten billion dollar Mackenzie Valley Pipeline Project of Canadian Arctic Gas required over $100 million of research and planning for their application. On a two per cent tax basis the government would have a fund of two million dollars for public interest research on the pipeline project which would be available to academics and outside consultants. These funds could be administered through the Social Science Research Council, National Research Council, or Institute for Research on Public Policy. Many ministers who sympathise with this problem of finance are still scared about questions in the House of

Commons[10] if the funding is to be directly from the Crown. The system I have described meets their problems and those of the public interest sector by providing an automatic flow of funds. Without such a change the public sector will have to continue on the basis of "ad hoc" research which is certainly not adequate from a qualitative or quantitative standpoint.

In the summer of 1975 the Cabinet began consideration of liberalizing the regulations on government documents so that the public would have a greater appreciation "about the background to the problems that the government faces." This is a basic change from the traditional approach of releasing only data which contributes to the public relations work of the ministry concerned. It would appear to be recognition of the principle of the public's right to know. It appears that the Trudeau Cabinet is prepared to consider that "research and analysis done internally through research contracts, or by consultants, be made more generally available to the public."[11] How far these changes may go is an open question, but after considerable lobbying from the public interest sector we hope that they will be substantial.

The financial problems of the public interest groups are well known to government and the principle of public funding has clearly been established by the Berger Commission of the Department of Indian and Northern Affairs. In addition, the Cabinet Committee on Social Policy considered a special memorandum on law reform at its meeting of June 13, 1975. One of the conclusions of that meeting was that the federal government has a responsibility "to provide for the representation of the public interest before regulatory boards." Yet, during this same period the Hon. Donald Macdonald, then Minister of Energy, Mines and Resources, refused to recognize the principle in any way in his negotiations with the Committee for an Independent Canada. Currently, the Department of Justice is preparing a policy paper "on an urgent basis" on "public representation before federal regulatory boards" but there have been no consultations with the public interest groups concerned to assess their problems.[12]

The Cabinet has clearly recognized the need for action to meet the real problems of the public interest sector; however I am rather suspicious that they will only agree to token actions in this direction.

In the winter of 1975-76, Marshall Crowe, the Chairman of the NEB, was challenged legally by three public interest groups for his earlier involvement as President of the Canada Development Corporation in the early planning of Canadian Arctic Gas. As a result of the Supreme Court finding "reasonable apprehension of bias", Marshall Crowe had to disqualify himself from chairing and sitting on the panel considering the

Mackenzie Valley Pipeline applications.

The irony of this action was that Crowe was certainly one of the most sympathetic figures in Ottawa; but the case was an exceedingly important precedent concerning all appointments to all government regulatory boards. The challenge was not to Crowe personally but to the arrogant way in which the Trudeau government acted with appointments such as the commissioners to the Bryce Commission. This will be the first of many such challenges until the government takes seriously the background of appointees. In addition, all such appointees must be excluded from later accepting post-retirement corporate appointments. Hopes for future lucrative corporate posts are a serious impediment to the objectivity of senior civil servants in Ottawa. However, some have not waited until they have left government service. Mr. Donald J. Morgan, Acting Director of Supply, Energy Allocations Board of the NEB, was forced to resign in 1974 when the CBC learned that a private company he had formed, Trident Trading and Transportation, had been buying and selling petroleum between the West Indies and the U.S. Mr. Morgan resisted pressure and claimed in public that his actions constituted no conflict of interest.

Currently both government and industry are reasonably happy with the status quo. A vigilant public interest sector would continually strengthen opposition questions in parliament so the multinational oil companies are desperately trying to keep the lid on the flow of public information for fear it will fuel the demand for public ownership. The NEB neatly filled its role in this scenario when Marshall Crowe opposed public funding for the public interest intervenors in spite of the precedents from the Berger Inquiry and the Porter Commission in Ontario.

Both industry and government, however, are in a period of serious change and confused relations. In the energy field the very nature of private capitalism is changing. With the great capital demands for future projects the multinational corporations are desperately hoping to get the necessary funds from increased consumer prices and tax concessions. This would allow the current foreign owners to stay in control for Canadian money would be expanding foreign ownership. Let the public take the risks and Exxon will take the profits. But even this was not enough in the Syncrude project and the governments had to be taken in as minority partners. This will be the first of many such partnerships through the Canada Development Corporation and Petro-Can. The federal government will have a steadily increasing stake in the oil business with its present free enterprise pattern of priorities and policies. This will erode further their ability to act as honest brokers to protect the public interest and consumer price levels.

There are many changes that must take place to reverse the present corporate ownership patterns. In the immediate term Canada must be prepared to subsidize informed and vigorous public interest intervenors and separate the regulatory functions of the NEB from its role as adviser to the government. Such changes would at least be a start to protecting the public interest.

1 Canada, House of Commons, *Debates,* 1957, 1155-60.

2 For analysis of the legal position of the NEB *see* Ian McDougall, "The Canadian National Energy Board: Economic Jurisprudence in the National Interest or Symbolic Reassurance," *Alberta Law Review,* XI, 1973, 327-82; *also* R.J. Gibbs *et al,* "A Review of the National Energy Board Policies and Practices and Recent Hearings," *Alberta Law Review,*IX, 1971, 535-6.

3 *National Energy Board Act,* 1959, Sec. 22 (1).

4 Eric Kierans, "The Day the Cabinet was Misled," *Canadian Forum,* March, 1974.

5 R.J. Gibbs *et al, op. cit.*

6 Edgar Dosman, *The National Interest,* Toronto: McClelland and Stewart, 1975, 27 ff.

7 Alberta Gas Trunk Line and Westcoast Transmission later withdrew from Canadian Arctic Gas to form the Foothills application.

8 Canada, House of Commons, *Debates,* 1971, 2614.

9 Ontario Energy Board, Rate Base Hearings on Equity Stock in the Mackenzie Valley Pipeline, Sept., 1975.

10 The Committee for an Independent Canada spent three years of exceedingly frustrating negotiations with the Hon. Donald Macdonald when he was Minister of Energy. While receiving vocal sympathy, the Cabinet refused any money.

11 The Government Priorities, Memorandum to the Cabinet from the Prime Minister, *Confidential,* Cabinet Doc. No. 465-75, July 22, 1975 (revised), Appendix A, Cabinet Committee Reports, 29-31.

12 Response to the Government's Priorities: Law Reform, Cabinet Doc. No. 379-75, June 3, 1975; The Government's Priorities, *ibid.,* 5-6.

1866
OIL EXCHANGE

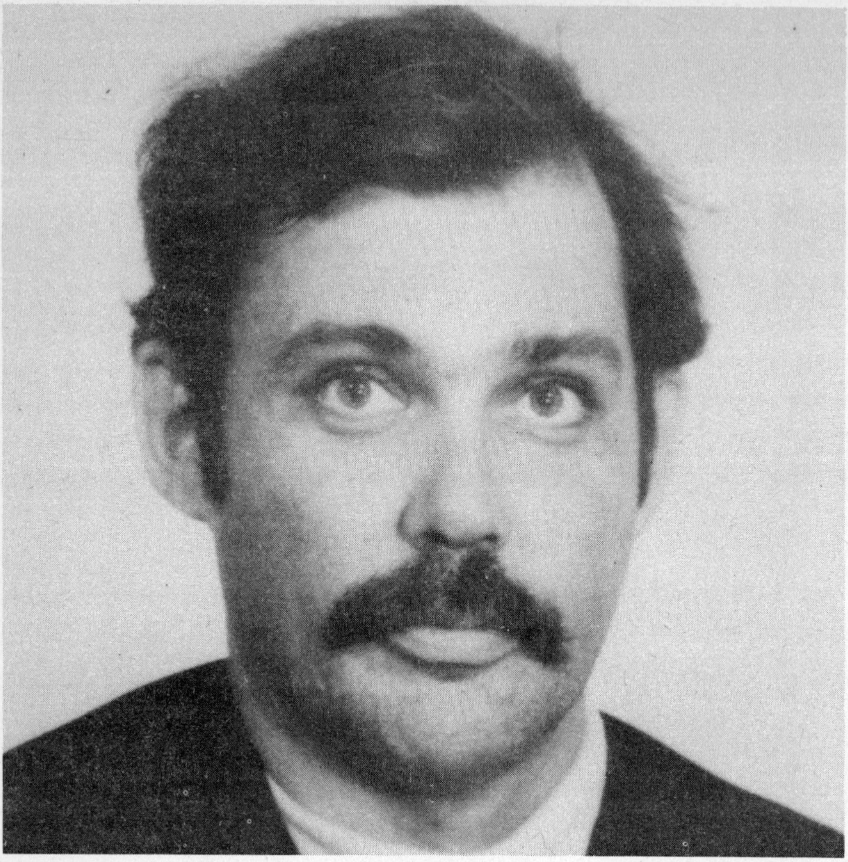

RICHARD GATHERCOLE

Richard Gathercole is an Associate Professor of Law at the University of Toronto and is Counsel to the Student Legal Aid Society at the University of Toronto. He is formerly the Executive Counsel in the Ontario Ministry of the Attorney General.

10

Richard Gathercole & James Laxer

And Now a Word from Our Sponsor

Few Canadians who today are old enough to have seen a hockey game on TV in 1960 will fail to remember Imperial Oil's famous "happy motoring" commercials with Murray Westgate.

Not a few Leaf fans can still remember the lyrics of the now nostalgic commercial:

> What a great, great feeling,
> What a wonderful sense,
> Of sheer enjoyment and of confidence,
> There's something you're aware of,
> Your car's been taken care of,
> At the sign of Happy Motoring,
> At the Happy Motoring sign.

When Murray Westgate, in his Esso dealer's uniform, saluted and told us in his gravelly voice to "always look to Imperial for the best", we thought of hockey, the wide open spaces of Canada, and the good life.

Today, Imperial's old commercials would no doubt highlight the fact for most Canadians that the price of gasoline has skyrocketed in the last few years. The old security they exuded about the good life in Canada under Imperial's protective care has now been replaced by anxiety about energy supply.

The new anxiety, just like the old security, has been coming to us carefully packaged by Imperial Oil. Today's Imperial ads play on our sense of uncertainty about Canada's energy future. Gone are Murray Westgate and the good life. In their place are scenes of oil rigs in the frozen Arctic and a new message:

Each year Imperial spends hundreds of millions of dollars on the big, tough, expensive job of developing petroleum supply and if Imperial is to continue to help Canada lessen its oil dependence on other countries, the company will have to put even more money to work.

The old messages were classics of what ad men call "image advertising". They sold us far more than Esso gas; they sold us a way of life and an image of Imperial Oil as a kindly protector and welcome symbol of security, available to us anywhere in Canada. The new messages are classics of a new form of advertising, "policy" or "advocacy" advertising. They have an urgent and active purpose. They are designed to influence current political debates about the price of gas, the appropriate level of profits oil companies should make, and about the exploration job the companies are doing.

Today the petroleum industry is the subject of daily political controversy and public criticism. "Policy" or "advocacy" advertising on television, radio and in periodicals to get the oil companies' story to the public is now a multi-million dollar business.

The Public Petroleum Association of Canada is a non-partisan public interest organization concerned with issues relevant to petroleum policy in Canada. It was created on the premise that the large foreign-owned petroleum companies have consistently dominated the formulation of Canada's energy policies—to the detriment of the national interest. It has sought, in its conferences, research and other activities, to provide a balance in debates about energy policy.

Operating with this perspective, the P.P.A.C. became concerned in the fall of 1975 with commercial messages being aired by Imperial Oil Limited on C.B.C.'s *Hockey Night in Canada.* The Association concluded that many of Imperial Oil's commercial messages were clear attempts to influence the current political debate about petroleum policy in Canada. To counter these, the Association prepared a one minute filmed commercial designed for use on *Hockey Night in Canada.* The commercial, shot at a cost of nearly $2000, was produced through volunteer efforts of professional film makers Clarke Mackey and John Clement and met the technical requirements for television.

The P.P.A.C. commercial showed a young father dropping his son off at a hockey arena. As the father drove home through Ontario countryside, past gas stations along the way, he thought to himself:

That kid—he really loves hockey.
Show him an Esso sign and he thinks of *Hockey Night in Canada.*
These days when I see an Esso sign I think of the high price of gas.

The oil companies say they need more money to find more oil.
That's how they justify the incredible profits they've been making.
And yet, it wasn't long ago that they told us this country had so much oil we should sell half of it to the Americans.
Foreign-owned oil companies telling us what's best for Canada—sometimes you'd almost think they put the oil in the ground.
Maybe it's about time we Canadians developed the oil and gas industry ourselves.

If you've had enough, let the government know you want the foreign-owned oil companies operated as a public utility.

The P.P.A.C.'s purpose in producing the commercial was to attempt to reach the *same audience* with its message that Imperial Oil was reaching. The Association did not believe it would be adequate to reply to Imperial Oil's controversial commercials on other programmes reaching different audiences.

On February 9, 1976, the P.P.A.C. wrote to Mr. Frank Selke, Director of Marketing and Promotion of Canadian Sports Network, to seek the purchase of one minute of commercial time on *Hockey Night in Canada.* The ensuing correspondence was the beginning of our education of how the definition of "advocacy" could shift about, depending on whether or not the "advocate" was a large corporation or just a small group of citizens with few resources.

February 9, 1976.

Mr. Frank Selke
Canadian Sports Network
415 Yonge St.
Toronto, Ontario

Dear Mr. Selke:

The Public Petroleum Association of Canada would like to buy one minute of air time on *Hockey Night in Canada* to show a commercial which we have prepared on the subject of the Canadian oil industry. ...

We are aware that *Hockey Night in Canada* is sold as a package and may be sold out. We feel it would be fair, however, for us to be able to buy one minute of time which is at present allocated to Imperial Oil, and we are of course prepared to pay according to the usual rate.

Our advertisement is ready for viewing and of course meets your technical requirements. We would be pleased to show it to you at your convenience.

We request that our ad be shown on *Hockey Night in Canada* on the southern Ontario network on one of the following dates:

Saturday, February 28, 1976
Saturday, March 13, 1976

We would be most grateful for a prompt reply.

Yours truly,

Mel Hurtig
James Laxer
Robert Page
Larry Pratt
John Richards
National Executive, P.P.A.C.

18 February, 1976

Dear Ms. Martin:

We wish to acknowledge your letter requesting purchase of one minute of commercial time on *Hockey Night in Canada* on the "southern Ontario network" either on Saturday February 28 or Saturday March 13. ...

As noted in your letter *Hockey Night in Canada* is sold as a package and commercial time purchased by the various advertisers is not subject to resale except at their request. At the present time the show is completely sold out.

You may wish to approach either the CBC, CTV or Global TV stations in southern Ontario who probably have larger amounts of time, some of which might be available to accommodate your regional requirement.

Yours very truly,

Frank D. Selke
Director, Marketing & Promotion

February 24, 1976

Mr. Al Johnson
President
Canadian Broadcasting Corp.
Ottawa, Ontario

Dear Mr. Johnson:

Enclosed please find copies of the Public Petroleum Association's correspondence with Mr. Frank Selke, Director of Marketing and Promotion of Canadian Sports Network Limited. ...

Mr. Selke's reply indicates that *Hockey Night in Canada* is completely sold out and that the only way commercial time can be purchased is at the request of the sponsors. In our letter to Mr. Selke, we indicated that even if the show is sold out, we believe it would be fair for us to buy one minute of time which is at present allocated to Imperial Oil. Moreover, the P.P.A.C. has no intention of itself approaching Imperial Oil to request one minute of its commercial time. We believe the responsibility for ensuring balance in the views expressed in commercial messages rests with the C.B.C., not with Imperial Oil.

We therefore request that the C.B.C. ensure that our one minute commercial will be aired on *Hockey Night in Canada* within the next few weeks.

Yours sincerely,

National Executive, P.P.A.C.

March 4, 1976

Mr. Albert Johnson
President
Canadian Broadcasting Corporation
Ottawa, Ontario

Dear Mr. Johnson:

...

We are most concerned that Imperial Oil is continuing its practice of presenting highly controversial political messages in place of product advertising. These messages outline the company's views on Canadian petroleum policy.

...

We believe that the C.B.C. must take ultimate responsibility for ensuring that a balance of views is allowed in matters of political controversy. This week, the nation's energy ministers are meeting in Ottawa to debate the price of oil. Imperial Oil is continuing to influence this crucial public debate through its ads.

May we ask you once again to let us know when our ad can be shown on *Hockey Night in Canada.*

Yours sincerely,

Anne Martin, Secretary

March 11th 1976

Ms. Anne Martin
Public Petroleum Association of Canada
212 Brookside Drive
Toronto, Ontario

Dear Ms. Martin,
...I note, from the correspondence you have enclosed, that Mr. Selke has indicated that this period is completely sold out. I am sure you will agree that it would be unfair, and unwarranted, to take back air-time from sponsors who meet the requirements of our Commercial Acceptance Policy.

Our Policy provides for time to be sold on CBC facilities to advertisers who have specific goods or services to offer and the content of their commercial messages must normally relate to the promotion of those goods and services. *Advertising messages of an institutional nature (image advertising) for such advertisers may present factual information about the advertiser's own projects and plans. Such advertising must not, however, urge CBC audiences to adopt a particular attitude or course of action in relation to these projects and plans. In our review of these messages we have found no attempt on the part of Imperial Oil to influence viewers' opinions in a manner which would contravene our Commercial Acceptance Policy.* (our emphasis)

All Imperial Oil messages currently scheduled on CBC *Hockey Night in Canada* have been reviewed and approved by the CBC Commercial Acceptance Dept. under these standards. In addition to promoting the sale of oil products and services, Imperial Oil does factually inform viewers in some of its messages of what it is doing or planning to do in the areas of oil exploration and development.

We entirely agree that "the CBC must take ultimate reponsibility for ensuring that a balance of views is allowed in matters of political controversy." And we respond to this responsibility within the context of our general programming where very clear policies determine the proper balance and exposition of all points of view.

With regard to your own proposed commercial, may I suggest that if the Public Petroleum Association of Canada wishes to purchase air-time in a period that is not already sold you should send your material to the Supervisor of Commercial Acceptance in CBC Toronto (and to Montreal as well if you desire air-time on our French-language network) where it can be reviewed under our Commercial Acceptance Policy.

Yours sincerely,

A.W. Johnson
President

In the italicized sentences of his letter, Mr. Johnson has provided a working definition of what can be called "product advertising" and "institutional advertising". It is our contention that in the last several years, a new type of advertising which we have called "policy advertising" has become more and more significant in the commercial messages presented to Canadian viewers. We contend that much of the advertising that is now labelled "institutional advertising" would more properly be called "policy advertising".

Policy advertising departs from institutional advertising through its intervention in the political arena on issues of political controversy. Policy advertising is directed at influencing the outcome of political de-

The Big Tough Expensive Job: Exploring for Oil

From the beginning of the oil industry in Old Ontario, Petrolia and Oil Springs were the Mecca for geologists and also for pretenders, ... "oil smellers" as these unorthodox professors and diviners are called ... Many thousands of dollars were spent by intelligent men in attempts to make a strike through the affinity supposed to exist between hazelwood and subterranean liquids.

There is no doubt that chance was responsible for the discovery of a good deal of oil ... The procedure was about the same in all cases. The first requisite was payment in advance to the "smeller" for his services ... He claimed the best results were obtained from a forked branch cut in the dark of the moon, with a knife, never before used, and that the sap must flow—otherwise the twig would not respond.

Victor Ross, *Petroleum in Canada,* 1917

bates that can be expected to affect the social, political and business climate of the advertiser. It is advertising whose purpose is to make that climate as much to the liking of the advertiser as possible. It allows powerful private corporations to influence political debates without having to subject themselves to the rough and tumble of having their views presented, along with the views of other citizens, to audiences of news and public affairs programming. Policy advertising bestows a unique privilege to communicate in the political arena on those with the ability to pay.

To date in Canada, this privilege is not balanced by the right of reply to the same audience by citizens who disagree with the content of policy advertising messages, even if they are willing to pay the full commercial rate for that right.

We readily agree with Mr. Johnson that Imperial Oil presents *some* commercial messages on C.B.C. *Hockey Night in Canada* that fit his definition of product and institutional advertising. In addition to these, however, Imperial Oil presents other commercial messages that, in our view, lie outside these bounds and are, in fact, examples of policy advertising as we have defined it.

Here is the text of an Imperial Oil commercial (aired in 1976 on C.B.C. *Hockey Night in Canada)* which we contend is not a proper institutional message, but instead constitutes policy advertising:

> At Imperial Oil bases and camps throughout the Arctic you'll find expeditors like Jerry Airy. They make sure that equipment and supplies get to their destination fast. Each year Imperial spends hundreds of millions of dollars on the big, tough, expensive job of developing petroleum supply. and if Imperial is to continue to help Canada lessen its oil dependence on other countries, the company will have to put even more money to work.

The message has the following politically controversial aspects:

1. The commercial uses the phrase "if Imperial is to continue to help Canada lessen its oil dependence on other countries . . . ." Many knowledgeable analysts of Canadian petroleum policy would maintain that Imperial's record has been anything but that of helping "Canada lessen its oil dependence on other countries . . . ." Instead, Imperial has been an unflagging advocate of the export of Canadian oil and natural gas to the United States, a policy that has led to the depletion of Canadian petroleum supply and which is now leading to greater dependence for Canadians on petroleum from other countries. Imperial Oil's commitment to exporting Canadian petroleum is well known and not disputed (certainly not by Imperial Oil itself). Two examples will suffice:

a) Imperial has made agreements for the export of much of the natural gas it has discovered to date in the Mackenzie Delta;

b) when J.A. Armstrong, now Chief Executive Officer of Imperial Oil, became President of the company he stated publicly that the company was operating on the assumption that Canada would harmonize its energy policy with the United States and would pool its energy resources with those of its neighbour to achieve a continental policy. (*The Globe and Mail,* July 9, 1970) While it may be natural enough for Imperial, a subsidiary of Exxon Corporation, to pursue continental, export-oriented policies, it is hardly consistent with the claim of their commercial message that the company is helping "Canada lessen its oil dependence on other countries..."

2. The commercial states that: "Each year Imperial spends hundreds of millions of dollars on the big, tough, expensive job of developing petroleum supply" and concludes that to be successful in this in the future "the company will have to put even more money to work." This section of the commercial is an obvious attempt to influence the central political debate on oil in the country—the issue of oil pricing. Since the Canadian price of oil was first frozen in September 1973, the petroleum industry has pressed for an end to the freeze and for the upward movement of Canadian oil prices to the world price. The industry has maintained that without higher prices and consequently, higher returns to the companies, there will be insufficient investment in the development of new Canadian petroleum supply. Examples of this sentiment among the oil companies are evident on the business pages of any newspaper. Two examples should suffice. In a letter to shareholders of the company on November 29, 1974, Imperial Chairman J.A. Armstrong complained about inadequate cash flow to the industry as a result of a federal budget. Armstrong concluded:

> The implications go far beyond the activities of Imperial Oil and the Canadian oil industry; they affect our country. If the tax objectives of the federal and provincial governments are not resolved in a way that lets the petroleum industry get on with the very large and very costly job of developing new reserves, Canada will find itself deficient in domestic petroleum supply.

The segment of the commercial cited above that refers to the company putting "even more money to work" is an attempt to influence this crucial political debate. It is by no means non-controversial to claim that the petroleum industry lacks sufficient profits to pursue successful exploration for new supplies. Prior to the world energy crisis of 1973, the Canadian petroleum industry predicted that it would be able to develop domestic supplies with far lower levels of cash flow than they now claim they need. This political issue was the subject matter of a meeting of the provincial premiers with the Prime Minister in Ottawa on May 6,

1976. Imperial Oil's commercial was attempting to influence that debate. The commercial has nothing in common with standard institutional advertising.

That Imperial's advertising is directed at influencing public policy is taken for granted in the business community in Canada. Speaking of Imperial's commercials at a marketing conference in Toronto in April, 1976, Peter Cameron, vice-president of Canadian Industries Ltd. of Montreal, said the ads were a model of "setting the record straight" campaigns. He said Imperial's commercials were needed because the media "all tell us how industrial marketers are contributing to the destruction of our environment, employing technological recklessness and reaping huge final profits."

In this frank discussion of Imperial's commercials, Mr. Cameron flatly contradicts the view in Mr. Johnson's letter to the P.P.A.C. that Imperial Oil's messages do not urge C.B.C. audiences to adopt a particular attitude in relation to the company's projects and plans. It is obvious to the business community and to those familiar with the current debate on energy policy, that Imperial's commercials are blatantly designed to influence C.B.C. audiences to view the companies' case for higher revenues with favour.

Imperial Oil itself admits to a motivation for its commercials that goes beyond Mr. Johnson's definition of institutional advertising. In an article by the Canadian Press on commercial messages, an unnamed Imperial Oil spokesman was quoted as saying:

> The company engages in corporate advertising because it believes that there is a public need for more accurate understanding and a balanced view of energy-related questions. (*The Globe and Mail*, April 22, 1976)

Clearly, Imperial Oil is attempting to influence C.B.C. audiences to take a view of the company's policy perspectives that suits the best interests of the company. Here is a second example:

> This is a report on the way Imperial Oil spent the money it received during 1975: operating costs accounted for 76¢ of each dollar received; taxes and royalties accounted for another 18¢. What was left? Six cents was what the company earned in profits. Part of these were paid to shareholders and the bulk of these earnings were reinvested in Canadian petroleum supply development.

This Imperial Oil commercial message attempts to do much the same thing as the first. By maintaining that the company earns a profit of only six cents on a dollar of revenue and reinvests the bulk of this profit in

Canadian petroleum supply development, Imperial is attempting to offset the public impression that the oil companies have been making very high profits and certainly have no need to make more. The public impression that the oil companies have been making very high profits is well founded in fact. Imperial's profits doubled between 1972 and 1974. When the company reports its rate of profit to its own shareholders, it uses the traditional method of calculating rates of return, that is profit as a proportion of shareholders equity. In 1974 this method of calculation showed Imperial's rate of profit to be 21 per cent (figure quoted from Imperial Oil's *Annual Report, 1974*), far higher than the essentially meaningless figure of six cents on each dollar of revenue. (The misleading nature of the commercial is that many people are unlikely to make the distinction between profit per sales dollar on the one hand, and the rate of profit on the other. Except in the case where they are both zero there is no necessary connection between the two. Six cents profit per dollar of sales is consistent with any rate of profit on capital. Thus, in 1974 Imperial's rate of profit was 21 per cent on invested capital, not six per cent.)

The commercial message is clearly designed to offset the public impression that oil companies have been making a lot of money. Offsetting this impression is seen by the companies as important in the current drive to achieve higher oil prices and higher revenues for the companies.

We have examined the texts of these Imperial Oil commercials in some detail not to argue that the company has no right to such opinions, but to demonstrate that these commercials deal with political matters. These commercials lie outside the bounds of institutional advertising and constitute perfect examples of what we have called policy advertising.

Our subsequent experience with the definition of "advocacy" as made by the private television stations was quite illuminating.

On April 6, 1976 Robert Cole, General Sales Manager of CFCF television in Montreal, contacted the Public Petroleum Association about the possibility of selling commercial time to the P.P.A.C. In his letter (which was not a response to an initiative by the P.P.A.C.) Mr. Cole stated:

> I have been made aware of your interest in airing a commercial which you have prepared on the subject of the Canadian Oil Industry.
>
> CFCF-TV in Montreal is privileged to have several oil companies as our clients and we would be pleased to sell you time on our station in similar vehicle used by oil companies.
>
> Mr. Doug Pearson, Vice-President of Alexander, Pearson, Dawson Inc., will contact you shortly to explain the position of our station in the market place.

Once contact had been established between the P.P.A.C. and Mr. Pearson, the offer to sell commercial time on CFCF was repeated. In a letter to the P.P.A.C. on May 6, 1976 Mr. Pearson clarified the offer to sell commercial time:

> ...CFCF-TV carried a limited number of Wednesday night Montreal Canadien hockey games which were jointly sponsored by Molson's Brewery, Imperial Oil and Ford of Canada. As a sponsored vehicle no spots were available to any advertisers on a selected basis.
>
> In furnishing other availabilities to you I cannot supply specific programs in which an oil company may have purchased a spot. As you can appreciate, any advertiser's campaign is considered confidential and it is against station policy to divulge information of this kind to anyone for whatever reason.

The purpose of the preparation of a commercial message by the Public Petroleum Association was specifically to respond to the "policy" advertising of the petroleum industry. It was our purpose to purchase commercial time on the same programmes on which such "policy" advertising was being broadcast. The CFCF offer to the P.P.A.C. did not fall into this category. It was a general offer to sell commercial time for the P.P.A.C. message.

After considering the CFCF offer, the P.P.A.C. decided to purchase commercial time on CFCF. In a letter of May 21, 1976, the P.P.A.C. informed Mr. Pearson:

> Further to your letter of May 6, the Public Petroleum Association of Canada would like to purchase commercial time on CFCF-TV as outlined in the schedule of availabilities that you have supplied.
>
> The Public Petroleum Association would like to buy one minute of time costing $280.00 on CFCF's Late Show programme within the next few weeks. This will be a one-time purchase only. We will determine whether more purchases are desirable after the first showing.

Mr. Pearson responded to this communication by advising the P.P.A.C. to submit its commercial message to the Telecaster Committee (the commercial acceptance committee of the private television stations including CTV network) so that it could be approved for broadcast. The P.P.A.C. followed this instruction and on June 2, 1976 received a letter from the Telecaster Committee which stated:

It is the policy of the Telecaster Committee to approve commercials which advertise goods and services and corporate good will. It is the opinion of the Telecaster Committee that advocacy advertising does not fall into any of these categories and therefore your commercial has been rejected on this basis.

In our view this experience with CFCF raises important issues that go beyond the concerns raised in the P.P.A.C.'s dealings with *Hockey Night In Canada*. CFCF first offered commercial time to the P.P.A.C. The Telecaster Committee then rejected the P.P.A.C.'s commercial on the sole ground that it amounted to "advocacy" advertising. Mr. Cole of CFCF-TV then made it clear to the P.P.A.C. in a telephone conversation on June 2, 1976 that his station was bound by the decision of the Telecaster Committee.

If the Telecaster Committee rejected the P.P.A.C.'s commercial because it consisted of "advocacy" advertising, then one would assume that CTV (the Committee's largest member) does not, at present, air any messages that can be described under that heading. Even a casual examination of the situation, however, reveals that this is far from the case. CTV carries many commercial messages that people in the corporate and advertising world commonly refer to as "advocacy" advertising. One example should suffice.

On May 12, 1976 *The Globe and Mail* reported an address by William Daniel, President of Shell Oil Ltd. on the subject of "advocacy" advertising. In his address Mr. Daniel announced that his company had temporarily abandoned plans to use its television commercials to present the corporate point of view. He stated: "There has been a real public backlash to some of this advocacy advertising."

CTV regularly carries the "advocacy" advertising messages of the oil companies, including the messages of Imperial Oil Ltd. on *Hockey Night in Canada*. Imperial Oil is a sponsor of the CTV national news where it regularly employs "advocacy" ads.

It appears, then, that the Telecaster Committee's refusal to approve the P.P.A.C. commercial does not arise out of any overall policy toward all advertisers. Rather it appears that the Committee refuses to approve commercials of public interest groups to air their "advocacy" ads while allowing corporations to use this kind of commercial freely.

In the case of the members of the Telecaster Committee, not only is it impossible for the P.P.A.C. to buy commercial time to reply to "policy" or "advocacy" ads of oil companies on the same programmes, but it is also impossible to buy time on any programme whether or not there is commercial time available.

This situation is surely intolerable. It means that equal access to the airwaves is being flagrantly denied by the Telecaster Committee in the way it applies its commercial acceptance policy.

What is the legal position? Who is responsible for this situation where there are different rules for corporations and for ordinary citizens?

Section 16(1) of the Broadcasting Act provides that:

> 16(1) In furtherance of its objects, the Commission on the recommendation of the Executive Committee may...
> (b) make regulations applicable to all persons holding broadcasting licences or to all persons holding broadcasting licences of one or more classes.
> (i) respecting standards of programs and the allocation of broadcasting time for the purpose of giving effect to paragraph 3(d)
> (ii) respecting the character of advertisements and the amount of time that may be devoted to advertising.

These paragraphs give a clear mandate to the Canadian Radio Television Commission (C.R.T.C.) to enact regulations outlining the nature of the obligation to provide programming balance and the authority to regulate the allocation of broadcasting time to ensure compliance with that obligation. Together with the authority to regulate the character of advertisements, this authority permits the Commission to direct a licensee to provide reply time similar to the United States Fairness Doctrine obligations for any commercial message or to ban policy commercials completely.

The right to reply on matters of public concern and the obligation of broadcasters to ensure this right are firmly established in the Broadcasting Act and in Commission policy statements, directives and decisions. Although the Commission has not considered the extension of the right to reply to commercial messages, it has often re-affirmed the proposition that broadcasters have a duty to provide programming balance and a reasonable opportunity for the expression of differing views on matters of public concern. "Programming" is defined in the TV Regulations as "anything that is broadcast" which clearly includes commercials.

In section 3 of the Broadcasting Act, parliament set out the principles on which Canadian broadcasting should operate:

> 3(b) the Canadian broadcasting system should be effectively owned and controlled by Canadians so as to safeguard, enrich and strengthen the cultural, political, social, and economic fabric of Canada.
> 3(c) all persons licensed to carry on broadcasting undertakings have a responsibility for programs they broadcast but the right to freedom of expression and the right of persons to receive programs, subject only to generally applicable statutes and regulations, is unquestioned.

3(d) the programming provided by the Canadian broadcasting system should be varied and comprehensive and should provide reasonable, balanced opportunity for the expression of differing views on matters of public concern and the programming provided by each broadcaster should be of high standard, using predominantly Canadian creative and other resources.

In the debate on this legislation in parliament on October 17, 1967, the then Secretary of State, Judy LaMarsh stated:

...the airwaves, which must be shared between public and private broadcasters, are public property and...they constitute a single broadcasting system. It is impossible to exaggerate the importance of broadcasting as a means of preserving and strengthening the cultural, social, political and economic fabric of Canada... It should be just as clear that every broadcaster, whether public or private, has a public responsibility for the effects of the programs he broadcasts. There must be reasonable opportunity for the expression of conflicting views on matters of public controversy. (*Hansard*, October 17, 1967, p.3174.)

Apart from the formal equal time rules of partisan political broadcasting, the Commission and its predecessor, the Board of Broadcast Governors, stated on a number of occasions its concern that broadcasters provide an opportunity for the expression of conflicting views on controversial issues of public concern.

In December, 1961 the Board of Broadcast Governors issued a policy statement dealing with controversial broadcasting. The policy of the Board in relation to controversial broadcasting was based on the following principle:

1. The air belongs to the people, who are entitled to hear the principal point of view on all questions of importance.

2. The air must not fall under the control of any individual or group influenced by reason of their wealth or special position.

3. The right to answer is inherent in the doctrine of free speech.

4. The full interchange of opinion is one of the principal safeguards of free institutions.

The foundation for these principles is the right to freedom of expression. The Board felt that this right can best be protected by "a policy which permits opportunity for the expression of varying points of view."

In an attempt to protect this right the Board promulgated a policy which provided for the:

Fullest use of the air for:

(a) forthright discussion of all controversial questions,
(b) equal and fair presentation of all main points of view,
(c) the discussion of current affairs and problems by informed, authoritative and competent speakers.

This policy formulation was given effect to by Ruling III of the same circular which placed on the licensee the obligation of providing broadcasting time to persons holding contrary opinions to those contained in a controversial programme or editorial previously broadcast. This policy circular remains the fundamental guideline of the Commission in the area of controversial broadcasting and its principles have been reiterated by the C.R.T.C. in a number of reports and Public Announcements including those involving the CBC programme *Air of Death*, the complaint by the Halifax Miles for Millions Committee against radio Station CHNS Halifax, and radio station CFCF Montreal's campaign against Bill 22, the Official Languages Act of the Province of Quebec.

These principles apply with even more force when the commercials in question are shown on, and time to reply is requested from, the C.B.C.

In the debates on the Broadcasting Act, the present Solicitor-General, Warren Allmand, noted the special position of the C.B.C.:

> ...this is a public network which is being supported by money of people from all across Canada... Someone, whether it is the government or the CBC administration must see to it that the corporation fulfills the purposes for which it was founded. Those purposes must not be twisted or frustrated. This does not mean that only the opinions of the government and the party in power are to be put forward. All opinions ought to be put forward. Only in this way can the CBC achieve its purposes of bringing about national unity and an exchange of opinion, as well as presenting Canadian contributions. (*Hansard*, October 17, 1967, p.3193.)

At the present time no Commission policy statement deals with commercial messages which implicitly or explicitly raise controversial issues or express a point of view on a matter of public concern. No Commission policy statement deals with commercial messages which may or may not fall within the prohibition of false or misleading advertising but which are by their very nature improper. No Commission policy statement deals with commercial messages whose very use is a subject of controversy.

It seems imperative to us therefore that Canadians ask the C.R.T.C. to

HOCKEY NIGHT IN CANADA

live up to its mandate. The P.P.A.C., in a submission to the C.R.T.C., made the following short term and long term recommendations.

SHORT TERM RECOMMENDATION

The Commission should take immediate action to prohibit the airing of Imperial Oil's commercial messages that fit the definition of "policy advertising", pending the holding of a Commission public hearing on the matter of advertising policy. Imperial Oil's "policy" messages are improperly influencing the current national debate on the price of oil. Imperial Oil's "policy" commercials are, at present, unanswerable before the same audience. The commercials should be discontinued at once.

Should the Commission decide against an immediate prohibition of Imperial Oil's "policy" advertising on C.B.C. *Hockey Night in Canada*, the Commission should ensure that the P.P.A.C. commercial message will be aired on C.B.C. *Hockey Night in Canada*.

LONG TERM RECOMMENDATION

In our view the Commission has two essential choices it can make to govern the phenomenon of "policy" advertising. One choice would be to draw up guidelines for "policy" advertising, including guidelines governing the right of public interest groups and other citizens to reply to such commercial messages. The second choice would be to prohibit "policy" advertising altogether.

It is our opinion that the second choice, the prohibition of "policy" advertising, is the preferred course of action. We believe that the subject matter of "policy" advertising differs substantially from legitimate "institutional" advertising through its attempt to intervene directly in matters of political controversy. We believe that political controversy is best dealt with through the ample time provided for news and public affairs programming. On such programmes citizens with different perspectives on public issues can reach the public with their concerns. This method of reaching Canadian audiences should be adequate for all—public interest groups, corporations, and individual citizens. Allowing "policy" advertising to continue means allowing privileged access to Canadian audiences on matters of public controversy for those with the means to pay for that right. Large private corporations have an ability to pay that makes token the right to reply for public interest groups and individual citizens.

At present private corporations have the right to reach Canadians with their own political messages through "policy" advertisements. They even have the right to deduct the cost of such advertising from their taxable income as a business expense, thus ensuring that the public pays 50 per cent of the cost of such commercials through taxes foregone.

We submit that "policy" advertising contradicts the spirit of Canadian broadcasting policy as it has evolved over many decades. We believe that the policy of the Federal Communications Commission in the United States—commonly referred to as the Fairness Doctrine—is not the best solution to the problem and that the Commission should endeavour to develop an approach consistent with Canadian experience and broadcasting traditions.

The Board of Broadcast Governors stated in 1961 that the air waves belong to the public and this principle is

> not promoted by the sale of network time to individuals or commercial concerns for broadcasts of their opinions or propaganda. The principles can be furthered by the provision of free time to competent speakers to present, without let or hindrance, varying points of view on questions of the day. The best safeguard of freedom of discussion is a policy which permits opportunity for the expression of varying points of view.

As the late Leonard Brockington, the first chairman of the Board of Governors of the C.B.C., said as long ago as 1939:

> Above all there should be no preference for wealth. Freedom of speech is not for sale at fifty dollars a minute on the air.

No Room at the Top

DOCUMENTARY

The Canadian as Manager

INTERVIEW BY DAVE GODFREY
WITH BRUCE WILLSON, AUGUST, 1976

Bruce Willson reminds you of one of the boys you went to high school with, the one who was tall, a good basketball player as much because of his size as his grace, not a great charmer of girls, fascinated by math but cool about it and too good-natured to be called "brains", but the one who, somehow, snagged one of the bright young women and then was off to university and the engineers and you lost touch with him, except everybody was sure he was going to do very well. And he did. Not as a subject for fiction, except the social parodies you avoided writing. Not believing in Roth and Richler somehow.

The drama of Willson's life, in a political sense, didn't arise until he was fifty. Before you meet him, it's hard to see why any drama arose. He went to school in Edmonton, lived in the university area and took a B.Sc. in Civil Engineering in 1943, became a Lieutenant with the 13th Company Engineers Services and Works, then went to work as Distribution Engineer for Northwest Utilities Limited of Edmonton. In 11 years he was Vice-President and from 1962-1965 he was President of that firm and of Canadian Western Natural Gas Company Limited. Twenty years of success, for firms that are controlled by I.U. International, formerly of New York City, now of Philadelphia. Total profits in 1975: $82,181,000. Acknowledged profits from Canadian operations in 1975 approximated $20 million through its 77 per cent holding in Canadian Utilities Limited.

Then President of Canadian Bechtel until 1971. Then President of Union Gas until 1974.

But there are two differences to his life that the pattern does not reveal. One, he gave up the success when the success seemed to be running headlong against the public good. And two, he remained calm, assured, fully rational in explaining why he left and in moving into his new role as supporter of the public interest.

So that even when you first meet him you have a feeling that his explanations will be coherent. True, at the moment he lives in this large house in the midst of the Bayview Country Club and the living room looks out onto the green of the fifth hole and the furniture doesn't come from Eaton's College Street, but true also that nor the house nor the country club nor the artifacts of success in any way possess him. He is his own calm man and always has been. No driveway made out of asteroid-gravel for him.

He is obviously contented. "I have always looked on energy supplies as not being that different from water, and the environment, and electric power, and sewage lines and roads; they're what you build an economy *upon*. The country can't stand the luxury of certain companies ripping off the consumer at prices related to what Arabian countries can charge an energy-depleted Europe rather than to cost. There used to be toll roads on Yonge Street and that may have been acceptable, as long as what they charged was related to maintenance and cost of capital. But not if the tolls brought in ten times the cost of the road every year."

He has taken a broad view and seen his own life in the context of the society and it has obviously brought a certain peace to him. Behind the upward drive to success lies a continuing conflict of conscience. Even in the early 60's he was upset by the studies he saw and had made of the long term hydrocarbon reserves available to Canada. And while a Director of the Canadian Petroleum Association (Alberta Division) he saw how difficult it was for many to link their interests as Canadians with the interests of their company. Even then there were doubts in his mind about the nature of oil industry propaganda.

But with Bechtel he was engaged in seemingly neutral activity. After the oil was found and after the producers had conformed to government policy, pipelines had to be built to carry the fuel to its destined market.

"It's not so much the profit motive that has to be called into question," he says, "as the cultural pressures for increased sales and profit every year. When you're dealing with a non-renewable resource there are only two ways to do that if your discovery rate is not ahead of your production rate: either sell more or sell it at a higher price. And both courses may be in complete conflict with the needs of the country. Over the years Bechtel played a major role in building Interprovincial Pipeline, Trans Mountain, TransCanada and West Coast Transmission, all of which have major export components. But just for an example of how your duties as an officer of a corporation can conflict with your responsibilities as a Canadian. In 1969 Bechtel was pushing to have yet another gas pipeline developed from north-east British Columbia and the Yukon to

California. From my discussions with industry people, I knew the reserves weren't there—no matter how minimal the protection you were to give to Canadian interests. But Bechtel wanted to have the new line developed, the California market needed additional supplies, the senior officers were in San Francisco and to keep their business volume up, to keep their profit picture improving, they were anxious to see it go ahead.

"I suppose in many ways my contrary views on this project confirmed that I was not suited to be a promoter of the additional export of Canadian energy resources. My resignation had been in the works for a year when I heard from Union Gas."

There, as a westerner, he found himself at a different end of the situation. The consumer's end. Union Gas holds franchises from Oakville to Sarnia, from Lake Erie to Georgian Bay. They are, in essence, a regulated utility; their profits are controlled. But not their costs. Of total energy use in Ontario, gas and hydro are about equal with 22%, coal now supplies only 12% and over 40% comes from petroleum. Founded in 1911 as an amalgamation of various south-western Ontario companies, Union Gas now only produces 3 to 4% of their supplies, another 6% comes from The United States, (under a contract terminating in November 1976), and 90% comes from Alberta. One of the first things Bruce Willson did as President was to update the studies of proven gas reserves and their relationship to Canadian needs. Looking at the charts that were available to him as President of Union Gas in 1971, it is no wonder that he took a strong public stance. Perhaps 40% of Ontario's gas-using industries obtained their supplies from Union Gas and Union Gas in turn obtained 90% of their supplies from Alberta. The availability of his native province's reserves thus had a direct and crucial bearing on the survival of Ontario's economy.

But his charts (1) showed a clear gap between reserves after exports and Canadian requirements, beginning early in the 70's and reaching crisis proportions by the early 1980's. Nonetheless, in 1970, the National Energy Board had approved an additional permit for export of 6 trillion feet of natural gas. A similar application was before them in 1971 for about 4 trillion feet with every expectation of similar approval. In total, Canada had signed away 18 trillion feet on 15 to 20 years contracts.

In the United States, even American sources clearly demonstrated that their deficiency would arrive even earlier (Chart 2) and be more extreme.

"It was fairly clear, even then, that the government was being sold a bill of goods by the oil companies. By 1980, despite sizable imports of one trillion cubic feet annually from Canada, American supply was going to be only 60 per cent of estimated market requirements. Yet, U.S.-owned companies in Canada were asking to export more and more of our

CHART 1

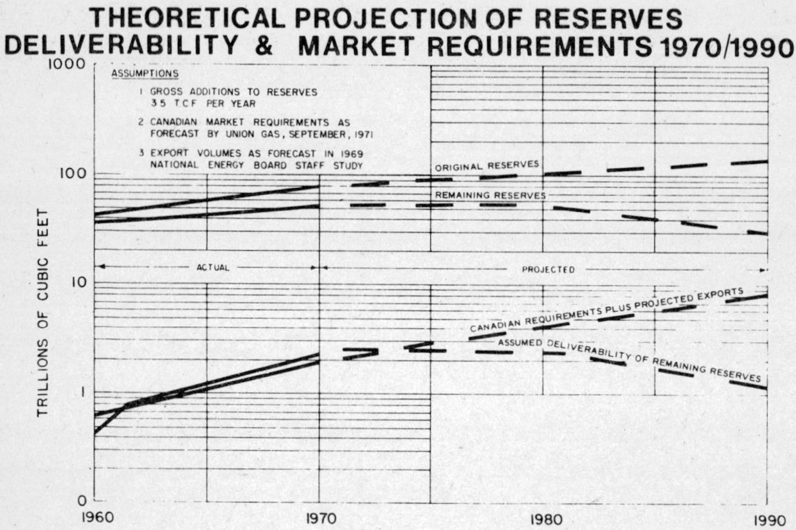
THEORETICAL PROJECTION OF RESERVES
DELIVERABILITY & MARKET REQUIREMENTS 1970/1990
ASSUMPTIONS
1 GROSS ADDITIONS TO RESERVES 3.5 T.C.F. PER YEAR
2 CANADIAN MARKET REQUIREMENTS AS FORECAST BY UNION GAS, SEPTEMBER, 1971
3 EXPORT VOLUMES AS FORECAST IN 1969 NATIONAL ENERGY BOARD STAFF STUDY
ORIGINAL RESERVES
REMAINING RESERVES
ACTUAL
PROJECTED
CANADIAN REQUIREMENTS PLUS PROJECTED EXPORTS
ASSUMED DELIVERABILITY OF REMAINING RESERVES
TRILLIONS OF CUBIC FEET
1000
100
10
1
0
1960
1970
1980
1990

CHART 2

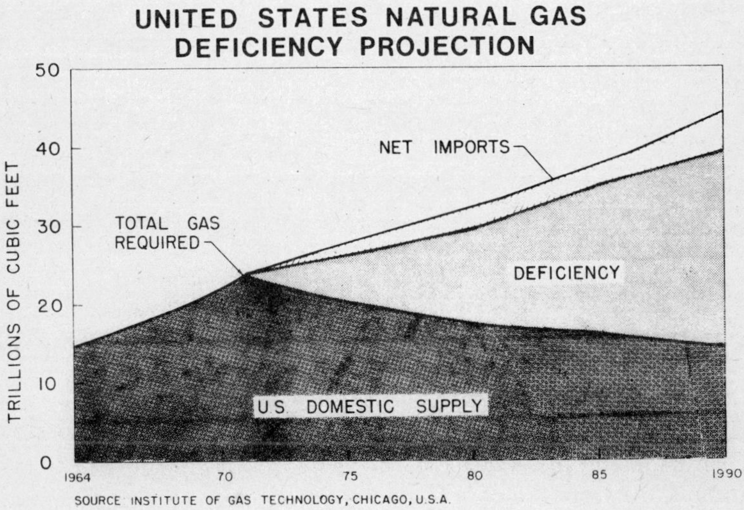
UNITED STATES NATURAL GAS
DEFICIENCY PROJECTION
50
40
30
20
10
0
TRILLIONS OF CUBIC FEET
NET IMPORTS
TOTAL GAS
REQUIRED
DEFICIENCY
U.S. DOMESTIC SUPPLY
1964
70
75
80
85
1990
SOURCE INSTITUTE OF GAS TECHNOLOGY, CHICAGO, U.S.A.

proven reserves and justifying this by speaking of vast potential reserves. Rather than the 53 trillion cubic feet of proven reserves, they were encouraging the government to think in terms of potential Canadian reserves of 725 trillion cubic feet. It was obvious they wanted the gas moved south as quickly as possible, to meet their rapidly growing deficiency.

"And very little thought was given to the cost of finding those 'potential' reserves and bringing them to market. The situation wasn't too difficult to understand. Between 1956 and 1970, proven reserves went up by about 7.5 per cent a year. Production went up at a rate of 16.3 per cent. That's a collision course you don't need a doctorate in math to understand.

"What we found, very early, at Union Gas, was that the government was operating on a theoretical basis and the market place was operating on the realities. Based on Ontario's rising use of natural gas, we knew we would need another 100 billion cubic feet by 1974, on top of our existing market of 170 billion cubic feet. Our principal supplier, TransCanada PipeLines, advised us that there weren't sufficient gas reserves available to it in Alberta to meet a contracted volume that large. In the past five years there had been few discoveries and many contracts made with American export buyers. We could only protect our users to 1974, three years, but the government was still going to approve exports to buyers for California and mid-west U.S. markets on long term contracts.

"At first, we were the only Canadian distributor to intervene. But the others came around. The facts were quite clear."

By 1971 then, before groups such as the Gas and Petroleum Association of Ontario, Bruce Willson was laying out facts which are only now entering the public consciousness. He was not rewarded for it.

"At the time, I think, I was characterized as very conservative. Later on, although we won a few battles like helping to stop the 1971 export approval of another 4 trillion cubic feet, some of the reactions were a little less polite. Some major figures in the oil industry felt I was undermining the whole free enterprise system. I was seen as a "prophet of gloom" when even Imperial Oil's *Annual Report*, in 1973, was speaking of hundreds of years' supply.

"I became more and more unhappy about three factors: the willingness of many producing companies to endorse, or at least not deny, estimates of reserves which bore little relationship to reality and at the same time, to expect, or advocate, rates of profit on existing production which were, in my view, astronomical. And to do this without any regard for the well-being of Canadian manufacturing and the public generally.

"Costs on the industry's inventory of reserves weren't going up by that much. These were the kinds of charts I was looking at (3 and 4) knowing full well that a lot of the gas had been found and developed at $.05 a thousand cubic foot and sold profitably and risen to about $.18 over a twenty year period, with profits being reported by most companies every year. And now suddenly we were talking not only about these large exports but about a ten-fold price increase by 1980 and about a wellhead price of over $8.00 by 1995.

"I just couldn't see myself participating in an industry that advocated such horrendous price increases, increases that would have such an adverse effect on the Canadian economy and on the cost-of-living for so many people. I couldn't."

This is the key question I must ask him. "But why were you the only one who quit? Why aren't there more Bruce Willson's? If anyone with any brains at the top of the industry knew what was going on quite early, quite a while before the government began to catch on to the industry's planning, why were you the only one who quit? Why didn't others?"

Picking his words with care.

"It's a comfortable life," he says. "In my case, I've been lucky . . . in becoming somewhat financially independent. So many are employees and they're beholden, for mortgages and expenses and their children. And stock options and pension plans. And relationships with major customers. I've been lucky. By 1974 our three children had at least one degree. I was working six and a half days a week, for people that, in some cases, didn't seem to appreciate what I was doing; I felt some of our directors didn't share my views on the responsibilities to the public of a company's management. There's so much pressure to report bigger and better profits every year. I felt I was just working too hard; occasionally I'd get up from the desk and feel dizzy. My decision to resign was fairly sudden when it came. I'd asked for something specific and the Board had turned it down. It was at the policy level. Most people tend to hang in when they're refused, I suppose. I talked it over with my wife Joan and wrote my letter. Now I'm doing a lot of reading and some writing and looking after my personal affairs. But my main interest is in helping public interest groups concerned with the energy problem. It was six months before the directors acted on my resignation.

"Then I was involved briefly with Syncrude. In April of 1975 Donald Macdonald asked me to be the federal government representative on the Board of Directors. I accepted. It's an example, I suppose, of the difficulties of communication. I went into the hospital for an operation and the government heard that I was at death's door. They asked me to stay on as

CHART 3

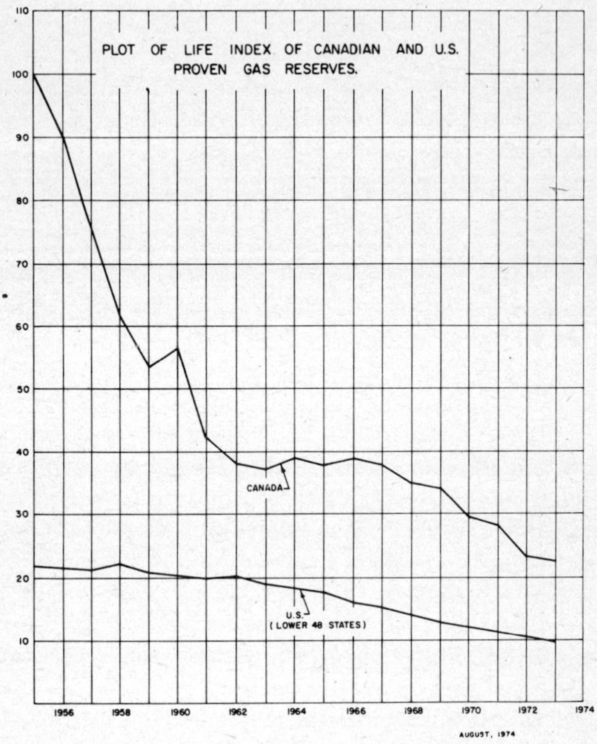
PLOT OF LIFE INDEX OF CANADIAN AND U.S.
PROVEN GAS RESERVES.
110
100
90
80
70
60
50
40
30
20
10
CANADA
U.S.
(LOWER 48 STATES)
1956
1958
1960
1962
1964
1966
1968
1970
1972
1974
AUGUST, 1974

CHART 4

PLOT OF ALBERTA WELLHEAD GAS PRICES
AS PROPOSED BY PRODUCING INDUSTRY

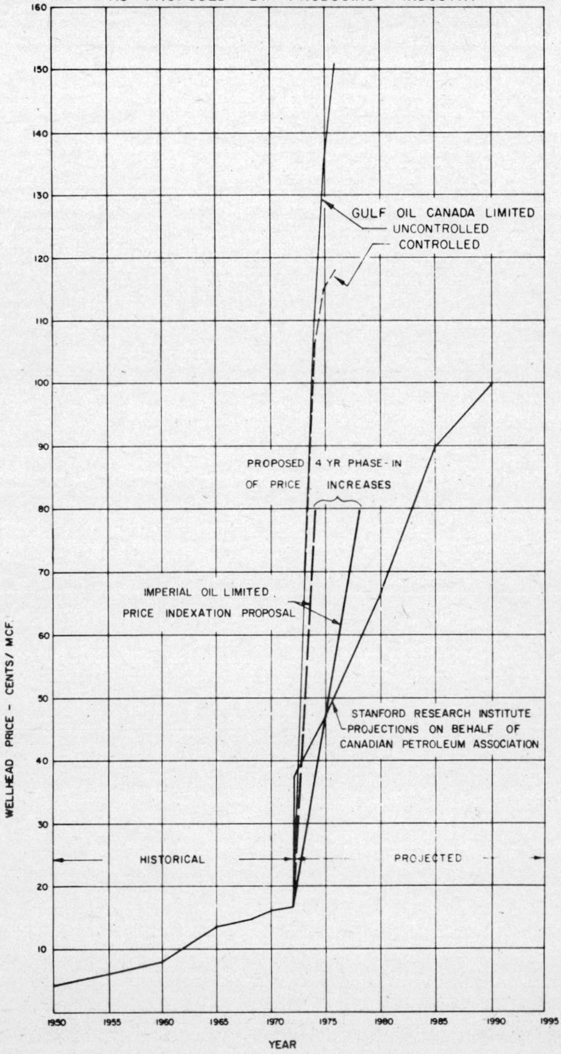

JANUARY, 1974

a Senior Consultant. But I felt that if they had changed their mind by May then I was better off to retain my independence. You hear, of course, that the oil companies had complained of my appointment. Whether this is true I have no way of knowing.

"But I basically feel we are running out of time to start rectifying our energy planning mistakes. We can still minimize a number of the upcoming dislocations if people lower their expectations a little, but the opportunity is slipping away. But it's possible; in the paper I gave to the Public Petroleum Association last November I outline how it might be done. If we start. And if we recognize that the U.S. control of the gas industry, by formal and informal means, must be overcome. It's not just through ownership of the majority of the shares of the producers, but in so many other ways. In the U.S. business works closely with government to help solve their energy problems, from an American point of view. We have to do that up here; we can't really expect them to do it for us. But it took us five or six weeks to get to see Macdonald and then you wondered if he was aware of the energy problems confronting Canada. As people, as individuals, I find the Americans delightful, capable and straightforward. But they're organized to have clout. They recognize that research and technological innovations are a key instrument of foreign policy, for example.

"And in this industry, the zipper effect doesn't always work to Canada's advantage."

"What's that?"

"Well, it's an American invention the zipper. But it means that, as part of corporate strategy, it's part of your job to establish a one to one relationship with your counterpart, in government, in other oil companies, and in related firms. It starts with the President, who deals with the Minister, and it goes right down the line, through the Deputies to the Chief in charge of collecting data. Zip. Zip. Zip. And that's why all the teeth mesh and forecasts tend to look so similar.

"Nobody but Canadians are going to protect this country's self-interest, that's what it comes down to. There's no way we're going to be able to help the Third World if we're flat on our backs ourselves. So much of our industry depends on energy; and at one time we had enough so that we had a real potential competitive edge, enough to make up for our higher wages, and higher heating and construction costs, and the smaller production runs that are inherent in the size of our market.

"But to throw that away, to take the so-called economic rent that the OPEC cartel made possible and split it up among two provinces, the American oil companies and the federal government, I just couldn't agree that that made any sense for Canada. I may have been a prophet of gloom, and more than one said I was, but I'm content to stand on the record."

It's hard not to agree with him. The coherence that you expected is there. On the walls of the living room, among the garden flower and fishing village paintings are some early pen and water-colour drawings of George Heath MacDonald. He, it turns out, besides being an architect, in 1959 wrote a history of Edmonton. In terms of national awakening, he may have been a little ahead of his time. But he surely would have been proud of his son-in-law, even if he too had been a little ahead of his time.

"One last thing," Bruce Willson says, "which I think bears out what we've been talking about. If you look now at the government studies what you will notice is that they are becoming more and more short term. We used to do forty years, but now we're down to twenty-five, twenty even fifteen years. And that's quite inadequate in terms of the job that must be done. That's not enough time for my children to raise their children. But if the government did a forty year study now, and released it, then it would show all too clearly the nature of the disaster facing Canada. And the reasons for it. But forty years isn't a very long time either, is it?"

TEXACO WARRIOR

4

PETROLEUM POLICY ALTERNATIVES

BRUCE WILLSON

Bruce Willson was born in Edmonton and spent most of his career in the gas industry in western Canada. He was President of Canadian Bechtel Ltd. from 1965-71 and President and Chief Executive Officer of Union Gas Ltd. (Chatham), 1971-74. Bruce Willson is a Director and Past President of the Canadian Gas Association and has served as a Director of the Canadian Petroleum Association (Alberta Division) and the Alberta Heart Foundation. He was Canadian Representative, International Gas Union Council, 1967-70. He is a member of the National Club (Toronto), the Edmonton Club, the Calgary Golf and Country Club, and the Royal and Ancient Golf Club of St. Andrews.

11

Bruce Willson

An Alternative Petroleum Policy for Canada

It is reasonably well known that the United States faces a gigantic energy crisis although its timing is believed by most to be some distance off. Not as well known is that natural gas shortages in the winter of 1976 were expected to cause large employment lay-offs, particularly in the eastern half of the U.S., with factories being unable to obtain essential fuel supplies. Interstate Pipeline companies anticipated being able to supply only 69% of the projected firm requirements of their customers. The shortages are growing annually. The 1976 deficit is expected to be 30% larger than that of 1975. Fortunately the 1975/76 winter was a mild one with the result that a fuel supply crisis did not develop to the degree originally forecast.

What is happening in the United States is an example of how lack of planning and overdevelopment of resources can slowly and progressively envelop a nation in a shortage of an essential commodity and how complacently the average citizen accepts a deteriorating situation at least until his own welfare is threatened. It is also an example of how the traditional business system is not coping with the forces at work and how senior government has failed to come up with essential solutions.

Unfortunately, by the time the natural gas shortages reach a certain magnitude it will be too late to take effective measures to head off the crisis. No doubt this point in time has already passed. It seems inevitable that economic catastrophies and human suffering must result from the increasing energy shortages in the U.S., although one should not underestimate American ingenuity and enterprise together with that nation's economic and military might to tackle and surmount difficulties which arise.

But surely Canada, a country rich in potential resources, but lacking in economic power should have the widsom and foresight to manage its affairs in order to avoid many of the errors which the U.S. has made in the energy field. By and large these qualities have been lacking as the country proceeds ostrich-like into the last half of the 1970's.

The following is a brief list of some of the factors involved in the energy problems confronting Canada today.

—First of all, Canada has a standard of living which generates a high per capita national level of energy demand together with a heavy dependence on hydrocarbon fuels, that is, oil and natural gas.

—Proven reserves of conventional oil and gas, in the traditional producing areas, particularly those of crude oil, are diminishing rapidly.

—Heavy export demand from the U.S. for new supplies of Canadian oil and gas continues on top of an already high level of deliveries to that country, although it now appears that the U.S. is realizing the limited nature of Canadian supplies.

—There has been a lack of significant oil or gas discoveries during the past ten years in the area south of 60 N. latitude and drilling results from exploration activities to date in the frontier regions have been disappointing.

—Many existing and prospective gas utilizing industries (except for those in Alberta) are unable to obtain or be assured of adequate long term natural gas supplies, thus hampering their planning and job-creating abilities.

—Canada has a need to import large and growing amounts of crude oil to meet the requirements of eastern Canada, that is, Quebec and the Atlantic provinces.

—Based on an extrapolation of past trends, the outlook is for alarming shortages of oil and gas in Canada for both the middle and long term.

—Canadians are experiencing skyrocketing wellhead prices for the low cost hydrocarbon reserves discovered and developed years ago.

—There has been a marked decline in exploratory drilling by the producing industry in spite of greatly increased industry profits.

—Canada has an exploration and producing industry which is largely foreign controlled, possessing its own corporate objectives and an uncanny ability to achieve them.

—The federal government has been all too willing to accept producing industry projections, policies and promises and adopt them as its own.

—Canadians are exposed to a provincial government (Alberta) with strong control on energy production, a vested interest in high wellhead prices and no apparent concern about the effect of those prices on the rest of Canada, its people or its economy.

—There exists an Alberta-sponsored plan to extract the heavier hydro-

carbons from existing gas production streams and thereby reduce the already limited energy supply available to provinces east of Alberta.
—There is a shortage of trained engineering and construction personnel with skills in the technological and managerial areas required for massive energy resource development projects. This results in a need to import the big international engineering firms.
—Native land claims continue to be unresolved and complex social issues exist in the Northwest Territories.
—Canada has an inhospitable climate with difficult and perhaps insurmountable physical barriers in the frontier regions.

Fortunately, Canada does possess some offsets—or at least, potential offsets—to this rather formidable listing of essentially negative factors. But they are not being pursued with the vigour, cooperation and sense of purpose necessary to solve or head off the coming energy crisis. Clearly, major policy changes are needed if the ultimately disastrous course upon which we are now drifting is to be altered so that the future generations do not "freeze in the dark" —a fate which has been predicted by some. (This suggestion implies the existence of prevailing policy, itself an arguable proposition.)

Just as clearly, it would be most unwise for Canada to rely on a programme of energy imports in a world where demand continues at rates which will soon outstrip even the prolific Middle East and where prices are set by the greatest monopolistic cartel in history. Strict conservation measures will help by stretching out existing Canadian reserves. What is essential for Canada, however, is a carefully developed, positive, comprehensive, *energy supply plan* backed up by the appropriate policies, legislation and the will to create long term energy self-sufficiency.

In order to suggest a course of action, it is useful to examine a few of the trends and developments of the past decade in the demand, supply and pricing of oil and gas in Canada.

First, it is important to consider the demand side of the equation. Canadians seldom stop to think how dependent on hydrocarbon energy resources their economy has become in recent years. To illustrate, in 1950 Ontario derived 23% of its energy needs from oil and gas. By 1973, just 23 years later, this had increased to 64% of a much expanded total requirement. Of this 64%, 41% was supplied by products derived from crude oil, and 23% from natural gas.

In addition to meeting the steep increase in domestic consumption—partly through imports—the oil industry in Canada has pursued a policy of seeking greatly increased oil and gas exports. It succeeded so well in convincing our governments and regulatory authorities of the wisdom of maximizing exports that in 1970, the then federal Minister of Energy, Mines and Resources, the Honourable J.J. Greene, defended Canada's aggressive export policy by stating that gas exports which had just been increased could be further expanded "but only if increased shipments of Canadian oil are also permitted into U.S. markets."

In June of 1971 Mr. Greene assured Canadians that "at 1970 rates of production, [Canada's] reserves represent 923 years supply for oil and 392 years for gas." When asked about the source of his information, he referred to the producing industry, the Canadian Petroleum Committee on Geological Reserves and the government's own National Energy Board. Mr. Greene is now a member of the Senate.

Throughout the first half of the present decade the oil industry has, with considerable success, kept up the pressure for still further increases in exports. By 1973, Canada was exporting 58% of its oil production and 45% of its gas production. At long last, however, Ottawa planners and politicians began to realize that a problem existed although only two years ago the then federal energy minister Donald Macdonald assured Canadians that the country was not facing a long term energy crisis. Mr. Macdonald's then deputy minister, Mr. Jack Austin, is now also a member of the Senate.

It is difficult to imagine an energy policy course more disastrous for Canada than the one which has been followed in the past. Recently, however, the seriousness of the outlook appears to have been recognized. In November, 1975 the government announced a cutback reducing oil exports to about 500,000 barrels per day, as of January, 1976. It is to be hoped that this growing understanding by Ottawa is not coming too late.

As far as energy supply is concerned, much confusion exists in the minds of the public regarding this somewhat technical question.[1]

To illustrate that the supply situation is deteriorating, the rule-of-thumb life index method can be applied to statistical data contained in the most recent report of the Reserves Committee of the Canadian Petroleum Association. Life index method means taking the year-end reserves and dividing them by the current rate of production to obtain an index which gives a very approximate but somewhat overstated indication of the number of years of adequate supply remaining.

When this is done, the result shows that since 1966 the life index for crude oil has declined from 24.6 to 11.9, a drop of 52%. The similar index for natural gas has gone from 38.6 to 22.0, a decline of 43%. Unfortunately, this does not mean that there is a 12 year supply of oil and a 22 year supply of gas. Canada now has to import about 50% of its oil needs and for the past year and a half indigenous gas production has not been able to meet Canadian requirements along with the presently authorized export level of one trillion cubic feet per year.

What about pricing developments? Not only has the producing industry failed to manage Canadian crude oil and natural gas reserves in a way which protects the long term supply needs of Canada, but it and the Alberta government are continually pressing to saddle Canadians with so-called "world" cartel-fixed prices. Wellhead prices for crude oil produced in Alberta leapt from $2.85 to $8.00 per barrel from 1972 to 1975 while prices for natural gas have skyrocketed from $.18 to about $.90 per Mcf during that interval. The federal government has concurred in these administered price increases under great political pressure from Alberta.

When these price increases are applied to the inventory of proven reserves existing at December 31, 1974, (that is to 7 billion barrels of oil and 53 trillion cubic feet of gas) it works out that the price to be paid for the remaining reserves has escalated by $70 billion. So the stakes in the game are huge.[2]

In the case of natural gas, these potentially crippling wellhead prices are minor compared with what some producer spokesmen have advocated Canadians should pay in the future. In February, 1975 a witness apearing on behalf of the Canadian Petroleum Association before the National Energy Board recommended that the field price of natural gas in Alberta should escalate further to $3.39 per Mcf by 1980 and to $8.04 per Mcf by 1995. Imperial Oil's separate gas price recommendation for 1980 was $3.54 per Mcf. It should be kept in mind that the price in 1972 was $.18 per Mcf. The annual cost of heating an average-size home in Ontario with natural gas would increase from $200 in 1972 to $2,000 in 1995 if these producer projections are achieved. The injustices for those on fixed incomes and the additional hardships for the poor that this will bring about are very serious. Also, the impact on such energy prices on Canadian manufacturing costs cannot help but be calamitous.

Yet, in its voluminous submission to the NEB, the Canadian Petroleum Association developed the historical average cost for the exploration and development of all oil and gas discovered by the industry during the period 1947-1973 as the equivalent of $.08 per Mcf. We have, then, an industry with the full assistance and cooperation of the Alberta government and now with the concurrence of the federal government, advocating the sale of a Canadian resource to the Canadian people at prices ten times to one hundred times original cost.

Is there any wonder that Canada has one of the highest rates of inflation in the world? In the light of such pricing policies, it is not surprising that natural gas rates are lower in the U.S. than they are in

Canada although it is the profligate use of energy in the U.S. which has triggered the shortages in Canada.

The producers' claim is that higher prices are necessary to fund the stepped-up exploration activity required to assure Canada's future energy supplies. Industry spokesmen make veiled threats about cutting back exploration effort if further increases are not forthcoming. An examination of a limited sample of industry data is useful in an attempt to understand and substantiate the legitimacy of the claim.

At the outset, it should be recognized that it it is very difficult to obtain comprehensive industry statistics because most of the major corporations engaged in exploration and producing operations in Canada either do not report the profitability of these operations separately from their other activities or the corporation is a wholly-owned subsidiary of a foreign parent and therefore does not report publicly its financial results. However, the following information on the combined results of two sizable exploration companies is instructive: Hudson's Bay Oil and Gas Company Limited (which does report publicly), 53% owned by Continental Oil Co. of the United States; and Mobil Oil Canada Limited (which does not report publicly), 100% owned by Mobil Oil Corporation, also of the United States.

For the period 1972 to 1974 when wellhead prices were escalating rapidly but production levels remained about the same, the companies' combined revenues from Canadian operations went from $322 million to $604 million, their before tax (but after royalty) earnings from $87 million to $206 million, and their after tax profits from $69 million to $124 million. Expressed as percentages, revenues were up 87%, before tax earnings up 135% and profits up 79% during this two year interval. In the same period expenditures for exploration by those two companies including the drilling of dry holes (unproductive exploration) went from $39 million in 1972 to $38 million in 1974, a slight decrease of about 2%. Exploration expenditures, as a percentage of gross revenue, declined from 12% in 1972 to 6.4% in 1974.

From this admittedly small corporate sample it would appear that the significant increase in wellhead prices which occurred in this interval did not generate stepped up exploration; rather, the reverse was true. It would appear to be true on an industry-wide basis also because exploratory wells declined by 21% in 1974 as compared with 1973.

This trend of declining exploration was carried forward into 1975. *Oilweek* magazine reports that exploratory drilling in the first six months of 1975 as compared with the similar period in 1974 declined by 13% in terms of the number of wells completed and by 20% in terms of exploratory footage drilled. At the same time, the exploration companies which report financial results publicly, by and large posted sizable increases in earnings as both oil and gas prices continued their upward spiral. Hudson's Bay Oil and Gas Company's earnings were up a further 24% in the first nine months of 1975.

The producing industry's claim that higher wellhead prices generate increased exploratory effort does not appear to hold water. It would be highly dangerous if not suicidal for Canada to place its reliance for future energy supplies on such performance and at such cost.

One must seriously question whether a system based almost exclusively on the profit motive will adequately and effectively meet the needs of Canadian society.

Over the years the petroleum business in Canada has operated under what is called the market system. As long as supply was adequate, competition kept consumer prices in line and, in general, virtually all segments of the industry prospered. Some companies such as oil and gas transmission pipelines, gas distribution and electric utilities were regulated by government boards because of their monopoly situations in the marketplace. Basically, they did a good job in serving the public interest at reasonable cost.

However, since 1972 when the shortages some of us had forecast earlier began to appear and particularly since October 1973 when the Middle East countries for a short period embargoed delivery of oil to North America and the OPEC cartel quadrupled its selling prices, the energy supply and pricing outlook in Canada from the consumers point of view has deteriorated rapidly. Nevertheless, the public is continually being exhorted by the oil companies to let the "market system" work, to increase exports to the U.S. and let supply and demand set the price. It is easy to understand why the entities which control the production leases might advocate such a laissez faire approach. However, the question needs to be asked: "Is it in Canada's long term interest to watch its self-sufficiency slip further behind as demand continues to exceed supply and prices escalate beyond all reason?"

Adequate energy supplies are too essential to the wellbeing of Canadians generally and to the efficient functioning of the Canadian economy to be treated as just another commodity to be sold off as quickly as possible. The standard of living, including the maintenance of present levels of social programmes, is directly tied to an adequate supply of reasonably priced energy. Consumers will pay virtually anything to keep warm in our severe climate. But are they not entitled to rely on their senior government to keep them from being unduly exploited by the power groups of society?

Surely these questions answer themselves. It is obvious that the "mar-

Corporate Images

The federal government's planned sally into the petroleum industry via Petro-Can has been drawing impassioned objections from the private sector.

But if ever there was offered a compelling argument against those objections and in favor of precisely such government intervention in the marketing end of the oil business, it has been offered by Imperial Oil.

Imperial, it turns out, acting in connivance with its parent corporation, Exxon of the United States, managed to squirm through a loophole that enabled it to avoid paying Canadian taxes while hiking product prices for Canadian consumers ...

By weaseling its way out of paying taxes while at the same time hiking product prices, Imperial has given substantial moral ammunition to those subscribing to the 'corporate welfare bum' school of thought.

It is difficult, however, to feel much sympathy for the oil company. The damage that has been done to its image of corporate citizenship has been done by its own hand.

—*The Edmonton Journal,* editorial, May 30, 1975.

ket system" cannot by itself solve the complex problems of energy pricing and supply which threaten the very existence of the Canadian economy. Nor can Canada safely rely on imports to make up the growing shortfall. To do so would be to ignore the limited magnitude of crude oil reserves worldwide and the hazards inherent in placing so much of Canada's future in the hands of the Arab world, to say nothing of the burden such a policy would place on the balance of payments.

It is essential for those indigenous reserves which now remain to be carefully managed and allocated to their most efficient uses if Canada is to survive until it is capable of achieving and perpetuating self- sufficiency in energy supply.

What needs to be done? Nothing short of an all-out, carefully planned, well organized, government-industry cooperative effort will be essential if Canada is to be able to meet the challenges presented by the approaching energy crisis. Such an effort would have to depart from some of the traditional views of the rock-ribbed free enterprisers who have dominated the petroleum industry. It should be based on principles of sound planning with the motivation being one of service to our fellow man in both this and future generations rather than simply a maximization of profits. Profits—an essential ingredient of our system where saving and investment is the heart of our economy—will take care of themselves in a well-ordered society.

The following are specific suggestions for a reordering of Canadian energy policy.

1. The creation of a comprehensive but somewhat flexible energy supply plan setting forth a realizable programme to guide future specific energy supply projects and designed to achieve complete self-sufficiency for Canada, in a reasonable period of time. The plan would be based on the development of known resources, not on hoped for discoveries or industry press releases. As economic discoveries or developments occured in the future, the overall plan would be amended to take them into account.

2. The cutting off of Canadian energy exports as quickly as practicable in order to stretch out the rapidly dwindling conventional reserves and thus meet Canadian needs for as long as possible. The cost of imports is likely to become prohibitive when worldwide petroleum energy supply becomes scarce.

3. The adoption of a cost-based approach to energy pricing. Cost would include reasonable royalties for the lessors as well as fair returns on the cost of capital directly invested or expended in developing energy supply projects. Such an approach would ensure that the consuming sector of the economy would not have to pay exorbitant premiums for Canadian resources. At the same time, it would provide appropriate returns for owners and investors.

4. The allocation of a significant part of the billions of dollars difference between prices now being paid by Canadian consumers and the "cost-based" approach to energy pricing for the development of such known resources as the Athabasca tar sands, heavy oil deposits and coal reserves, thereby assuring their financing. The estimated difference of $70 billion is 35 times the forecast estimated capital cost of the Syncrude project so it is important to realize that there are tremendous sums of money potentially available for energy resource development projects.

5. The allocation of another significant part of this difference between today's prices and actual costs to major research and development projects aimed at such possibilities as in-situ production of bitumen from tar sands and gasification of coal.

6. The allocation of yet another part of this difference to conventional oil and gas exploration, primarily in the frontier areas, perhaps through Petro-Canada. In recent years the industry has threatened to withdraw its expertise unless its pricing demands are met. It is therefore essential that Canada possess its own technology and the ability to apply it.

This approach and these suggestions are tantamount to a planned economy for energy and anyone advocating them might well be accused of embracing socialism. But that is not necessarily the case. Canada has a long tradition of public initiative, public responsibility and public ownership in energy production. These are not radical concepts. Ontario Hydro, for example, is 65 years old and it is the offspring of a Conservative government.

The rapid depletion of non-renewable pretroleum resources coupled with the great dependence Canada's economy now places on oil and natural gas require a solution which the private enterprise system by itself, and with its high degree of foreign ownership, is not delivering. Strong central policies and direction are essential if the needs of the future are to be met in a timely fashion. Anything less than this will fall short of what is required to run the engines of the economy. In a crippled position, Canada will be unable to meet its own internal goals, let alone fulfill a significant role in assisting the less well-endowed nations of the world.

It would be desirable that the best features of the enterprise system be preserved and that entrepreneurship be properly rewarded. True competition—generally speaking—is a good thing. But when too much power is centralized in a few corporations or owners and that power is exercised to the detriment of society, then it is time for the national government to discharge its responsibilities to the nation as a whole.

There are many capable Canadians in the petroleum industry whose efforts need to be devoted primarily to assuring the future of their country rather than to adding a few cents per share to the worldwide earnings of their multinational parents or to swelling the coffers of the Alberta Heritage Fund. Let's dedicate ourselves to one of the great challenges facing Canada as a nation.

1 Frontier potential is excluded from this brief analysis because of its remote location and unknown magnitude—as well as the many other uncertainties surrounding its possible availability.

2 It should be remembered that only a portion—albeit a significant one—of the revenues realized from the skyrocketing prices accrues to the producer. While it is difficult, if not impossible, to obtain accurate information applicable to all situations, it appears that the largest share of the greatly increased prices goes as royalty payments to the producing provinces, primarily Alberta. The federal government, through the corporate income tax structure, receives a substantial increase in its revenues. The balance, perhaps 30 per cent, remains with the producer.

T.C. DOUGLAS

For more than thirty-five years Tommy Douglas has been in the forefront of Canadian politics; as Member of Parliament; as Premier of of Saskatchewan; and as National Leader of the New Democratic Party. He is the Member of Parliament for Nanaimo-Cowichan-the Islands. In the House of Commons he has been a skilful debater and has fought hard on issues such as federal medicare, economic independence, the Vietnam War, and the War Measures Act. In his role as energy critic for the New Democratic Party, Tommy Douglas has consistently warned Canadians against policies that have led to the current energy crisis.

12

T. C. Douglas

The Case for Public Ownership

Despite the constant discussion in Canada about a "national energy policy", the fact remains that Canadian energy priorities continue to be subordinated within continentalist policies. Canada's resource development policies have been formulated not by governments, but by the multinational oil and gas corporations operating in Canada. Ninety per cent of oil production and ninety-nine per cent of petroleum refining is owned and controlled by foreign, mainly American, oil interests.

Let me make clear what is meant by a continental energy policy. The United States has long recognized that its energy resources as well as its mineral and forest resources are dwindling rapidly. The U.S. has looked to Canada as a reservoir to be readily tapped. The continentalist approach is analagous to two young men who have gone out on their summer holidays together. One has spent most of his subsistence and he says to his friend who still has some money in his pocket, "let's pool our resources." That's the continental energy policy. And it should be remembered that the oil and gas companies who came to this country didn't come as an act of generosity. They came because the United States needed resources and Canada had resources which were profitable to extract.

Continentalism in petroleum policy was built into the very fabric of Canada's so-called National Oil Policy of 1961. The most important aspect of this policy, adopted by the Diefenbaker government, was the creation of a north-south line through the Ottawa Valley. Areas west of the line would use Canadian oil while areas to the east would import their oil. Vancouver, under the policy, could not import oil from Indonesia and Ontario could not import cheaper oil from the Atlantic seaboard. From 1961 to the time of the rise in the international price of oil in 1973, Ontario had to spend between $70 and $80 million more per year for its oil as a result of being required to puchase more expensive domestic crude oil. The area west of the Ottawa Valley was thus a captive market for western oil production which was almost entirely owned and controlled by American companies.

East of the Ottawa Valley, the market was also reserved for American oil interests. Imperial's parent company, Exxon, supplied this market with sour oil from Venezuela which it could not sell in the United States. It is significant that every refinery east of the Otawa Valley line had to be equipped with machinery to remove the sulphur from this Venezuelan oil.

The National Oil Policy was deliberately designed to create a surplus of western Canadian oil which would then be siphoned off to the United States. Canadians would do well to remember, when they are given depressing figures about our oil situation, that Canada exported as much as 60 per cent of its daily production to the United States year after year under the National Oil Policy of 1961.

And when some of us protested—inside the House of Commons and outside the House of Commons—about shipping out oil despite dwindling resources, our arguments were countered by the multinational oil companies who provided estimates to show that this country had abundant supplies and that we could afford to export our surpluses.

In 1970, the very same companies who presented statistics to the NEB and to the Cabinet to prove that Canada had enough gas supplies to last far into the next century as well as abundant oil reserves are now telling us that we are in a crisis situation. These companies are appearing before the government asking that they get special concessions to develop the oil sands and that they get the permit to build the Mackenzie Valley Pipeline in order to obtain gas.

Five years ago we had plenty. Now they tell us the bad news that we're running out of oil and gas. The tragedy is that the government of Canada ignored all the warnings. They accepted the estimates of an industry which has consistently misled and lied to the Canadian people and to the government and to the National Energy Board. The result is that Canada now has eight or ten years supply of oil left. This year, Canada, which has been a net exporting country, will have a net deficiency of 100,000 barrels per day. Next year the deficit will be 300,000 barrels per day. Rationing of natural gas is possible in the city of Toronto within six years.

What is equally serious is that every day there is a deficiency, Canada is becoming increasingly dependent upon imported oil. For compensation payments to the multinational oil companies who import the oil, Canada will pay out $1,685,999,000. Canada will be paying the multinational oil companies five dollars per barrel as the difference between the domestic and the international price for every barrel they bring into the country. When I see us paying out $1,685,999,000 and then hear people say "where will you get the money if you want to nationalize the oil industry?" I say we've already paid for the oil industry half a dozen times over.

Who is to blame for the plight we're in? There is no doubt who is to blame: the multinationals which have dominated the industry ever since they bought out the small Canadian firms; supine governments who have been completely subservient to the lobbyists of the international oil companies; and an indifferent public which has ignored all the warnings.

It is future generations who will have to pay the price of our resource mismanagement and our public apathy.

There is not much use complaining about the past. What matters is that these same multinational companies, having depleted our conventional western reserves, are now starting to put new words to an old tune. They've got new plans for a continental energy policy. Are the Canadian people going to buy them? Is the government going to buy them?

There's an old saying: "If they fool you once, shame on them. If they fool you twice, shame on you." The Canadian people are going to have to decide whether they will allow the continuing mismanagement of their resources.

There are two main programmes that the foreign-controlled oil companies are now pressing in Canada. The first is the development of the Athabasca oil sands. With the shortage of supply, the oil companies involved are now holding the Canadian government up to ransom, saying, "you want the oil sands developed? How much?" In a period of less than twelve months Imperial and Gulf, two of the participants, said that the estimated cost of the project had gone up from $980 million to over $2000 million. They claimed that they could not go ahead without substantial government input. And they just sat tight.

The federal, Ontario and Alberta governments subsequently put in large commitments for equity capital, for the power plant and for the pipeline. Furthermore, the federal government agreed to tax concessions and royalties that had never been offered to any other petroleum project.

And they got more than that. The consortium got a guarantee from the governments that the price of Syncrude oil would be the international price. In consultants' reports to the Energy Committee of the House of Commons, it became clear that the economic financing estimates for Syncrude were based on $15 per barrel prices when the oil

We pay but oil firms call the tune

The private industry itself has provided ample justification in the past two years for massive public involvement. When the federal government decided in 1973 that Canadian oil companies should not be allowed to reap windfall profits from prices dictated by the Organization of Petroleum Exporting Countries (OPEC), exploration dropped drastically in Canada.

This private business decision to cut back on exploration for new reserves came at a time when Canadians were suddenly discovering that existing supplies of oil and gas were rapidly running out. Yet we were helpless to do anything about it.

An industry survey indicates there will be a marked improvement in exploration activity next year, mainly because the federal and provincial governments involved have made important taxation and royalty concessions to the private companies. One industry spokesman says the renewed activity "is being undertaken on good faith that wellhead prices will go up and that the governments will further improve the royalty and tax position of the industry."

In other words, Canadians will get the oil and gas they need provided they allow the private (foreign-controlled) petroleum industry to charge what it likes, and so long as governments don't try to get too pushy.

This is an intolerable situation. And it should be the foundation stone of any new policy that Canada's energy future will not be left to the whims of the private industry. . . .

Petro-Canada, the new federal government oil company, is a step in the right direction. But it is sadly under-financed and, by former energy (now finance) minister Donald Macdonald's own instruction, under-ambitious.

In any case, according to a federal energy department study, it would take 15 years or more to build an effective, fully integrated petroleum company from the ground up. Canada's energy needs are much more urgent than that.

The way to deal Canadians in on the high-stakes game of energy development is to take over an existing company, lock, stock and ($8) barrel. The logical candidate to provide the vital public leverage is the industry leader, Imperial Oil Ltd.

—*The Toronto Star,* editorial, Dec 4, 1975.

comes on stream, going up steadily to $35 per barrel within the life of the plant.

The Syncrude project also got a concession from the government that Syncrude oil can be exported if it is surplus to Canadian needs.

Finally, the government agreed to raise the price of Canadian oil to the world price on or before the time Syncrude goes into operation.

In total, the people of Canada, through the governments of Ottawa, Ontario and Alberta, are putting up 70 per cent of the cost of Syncrude but are receiving only 30 per cent of the control. If we were led down the garden path before under the National Oil Policy, then I am saying now, under the Syncrude plan, Canadians will be giving away one of the greatest resources possessed by any country in the world. And they will be giving it away with their eyes open.

The second project of the new continental energy policy is the Mackenzie Valley Pipeline. The companies which are pushing this project are the same ones who five years ago campaigned across Canada and presented ficticious figures to the National Energy Board in order to increase the gas export limit by 50 per cent from 12 trillion cubic feet to 18 trillion cubic feet.

The argument for building a Mackenzie Valley Pipeline in the 1970's is that Canadians need Arctic gas. But the Mackenzie Valley Pipeline is a project whose primary purpose is to take Alaskan gas from Prudhoe Bay to the United States.

To justify the pipeline—even a small diameter pipeline—15-20 trillion cubic feet of gas must be proved. The highest estimates of Canadian reserves in the Delta area have been 4 and 5 trillion cubic feet. The Prudhoe Bay reserves are much larger but are combined with oil. Under Alaskan conservation law, the companies can only take out the gas as the oil is produced. The gas load will thus grow only very gradually. The idea is that Canadian gas be used to fill the pipeline in the early stages while the Alaskan load builds up. It will take only a few years for the 4 or 5 trillion cubic feet of Canadian gas to go through the pipeline. After that, the pipeline will be carrying almost exclusively the Prudhoe Bay gas. And we need not expect the Americans to sell any of that gas to us.

Even apart from environmental concerns and the land claims of native people, there is no economic justification for the Mackenzie Valley Pipeline. Canadians are being asked to spend $8 billion and to bear the price of the accompanying economic dislocation for the sole purpose of taking American gas to the United States. I am suggesting that we Canadians are being played for suckers if we proceed with the Mackenzie Valley Pipeline.

The time has come for Canadians to develop their energy resources in

Majority Favour Public Ownership

November, 1975
Canadians are facing rapidly rising prices for oil and gas as well as uncertainty concerning future energy supplies. In view of this do you think that the large foreign-owned oil and gas companies should or should not be nationalized by the Canadian government and operated as a public utility?

SHOULD 1
SHOULD NOT 2
QUALIFIED (SPECIFY) 3
DON'T KNOW 4

The Canadian Institute of Public Opinion, the organization which conducts the Gallup Public Opinion Polls in Canada, was commissioned by the Public Petroleum Association of Canada to conduct the study.

		Region				
	Total	Atlantic	Quebec	Ontario	Prairies	B.C.
TOTAL INTERVIEWS	1038 100.0	100 100.0	282 100.0	375 100.0	171 100.0	110 100.0
SHOULD	530 51.1	51 51.0	155 55.0	201 53.6	67 39.2	56 50.9
SHOULD NOT	318 30.6	31 31.0	77 27.3	103 27.5	69 40.4	38 34.5
QUALIFIED	16 1.5	—	5 1.8	6 1.6	2 1.2	3 2.7
DON'T KNOW	173 16.7	18 18.0	44 15.6	65 17.3	33 19.3	13 11.8
NOT STATED	1 0.1	—	1 0.4	—	—	—

	Sex		Occupation			
	Male	Female	Prof/ Exec	Sales /Cler	Labour	Other
TOTAL INTERVIEWS	517 100.0	521 100.0	237 100.0	144 100.0	433 100.0	224 100.0
SHOULD	268 51.8	262 50.3	114 48.1	72 50.0	230 53.1	114 50.9
SHOULD NOT	180 34.8	138 26.5	90 38.0	49 34.0	126 29.1	53 23.7
QUALIFIED	8 1.5	8 1.5	5 2.1	2 1.4	7 1.6	2 0.9
DON'T KNOW	60 11.6	113 21.7	28 11.8	21 14.6	69 15.9	55 24.6
NOT STATED	1 0.2	—	—	—	1 0.2	—

	Income			
	Under 6000	6000-9999	10000 14000	15000 & up
TOTAL INTERVIEWS	149 100.0	150 100.0	259 100.0	280 100.0
SHOULD	80 53.7	79 52.7	149 57.5	140 50.0
SHOULD NOT	35 23.5	37 24.7	69 26.6	111 39.6
QUALIFIED	2 1.3	1 0.7	5 1.9	4 1.4
DON'T KNOW	32 21.5	33 22.0	35 13.5	25 8.9
NOT STATED	—	—	1 0.4	—

the interests of the Canadian people.

The Gallup Poll of November 1975 sponsored by the Public Petroleum Association showed that 51 per cent of those interviewed favoured nationalizing the oil and gas industry. That is understandable. There has been a tremendous change in public attitude. People have given the great corporations their day. They've given them a free hand. They have seen them grow wealthy beyond their wildest dreams.

Therefore, I say that the answer lies in a programme of nationalization of our energy resources.

Those who favour nationalization should realize that the case is weakened if we fail to recognize the realities of the situation. Nationalization is not a magic cure-all. Nationalization will not put one barrel of oil into an empty well. Nationalization is an economic device by which the representatives of the people can plan the resources of the nation for the benefit of its citizens.

Canadians should be very wary about buying up depleted oil fields. Venezuela has recently paid $475 million to Exxon for an operation which had an original investment of $140 million and from which Exxon had already taken millions of barrels of oil. Britain had the problem of nationalizing coal mines which had almost entirely been depleted of coal.

We shouldn't become sentimental about the idea of nationalization. Nationalization simply provides the means to do the job. Nationalization is not completed by the simple act of taking over private corporations. Unless this is made clear to the public there will be expectations which will not and cannot be realized until a great deal of planning and work has been done. The Canadian people must realize that having taken fifty years to get into this awful mess, it is going to take quite a lot of time to get out of it.

Now what is the role of nationalization? How do you go about it? In my opinion the government should take as a first step the acquisition of one of the major oil companies. Imperial Oil is the logical choice because it is the largest and because it would give the federal government access to oil and gas leases which come under provincial jurisdiction. Most people forget that. Petro-Canada cannot go into the provincial oil fields. Provincial governments have jurisdiction. But by buying out, let us say Imperial Oil, access to oil and gas leases which come under provincial jurisdiction is immediately gained. It would bring oil and gas pipelines, refineries and distribution facilities under federal ownership.

What is the advantage of doing that? The first advantage is that the large profits which are now being siphoned off to the United States would now be available for further exploration.

The oil companies are not spending the increased profits gained from higher priced oil and gas on exploration. The argument put forward by the oil companies for raising the price is this: higher prices will assure higher revenues and it will then be feasible to go out and look for more oil. Remember that only a few years ago the price was $3 per barrel. The present price is $8 and the government has said that the price will continue to rise until it reaches the international price, probably $15 per barrel.

With the raising of the oil price, profits have skyrocketed. But has there been an increase in exploration? The answer is no. I have had a very careful research project done which shows that over the last number of years the expenditure for exploration as a percentage of the oil industry's profits has steadily declined. This is particularly true in the prairie provinces where oil could still be found both in the shallow pools and by using stepped-up recovery programmes. The oil companies have in fact been on a slowdown strike which they like to call a "loss of investment confidence".

But even if the increased cash flow had led to much-increased exploration, there is still an assumption made which is fundamentally wrong. The assumption is that Canadians should not only pay for the barrel of oil which they are consuming now, but should also pay the companies money to find the next barrel of oil which will belong to the companies. In the financial breakdown for producing a barrel of oil published by the Canadian Petroleum Association in 1975, the companies entered $1.55 for future exploration and development—$1.55 for every barrel for hunting for the next barrel. I say that if Canadians are going to pay to find the oil, the oil should belong to Canadians.

Nationalization would also enable Canadians to control the development of the oil sands and offshore oil and thus control the benefits from these undertakings. Canada would be better able to carry on proper conservation and an effective pacing of these programmes to meet domestic needs. Most important of all, a nationalized oil and gas industry would facilitate more effective handling of environmental problems, particularly in extracting oil from the Athabaska sands—problems which are tremendous.

A nationalized oil industry would also enable Canada to enter into state trading agreements with oil-producing countries. Venezuela and Iran have already indicated their willingness to enter into such agreements at lower prices than they are charging to the multinationals. The Shah of Iran said in 1974 that he is prepared to reduce the price if Iran is selling to other governments, but there is no use in reducing the price to Exxon because they would put any savings in their pocket rather than passing them on to the consumer.

AS OIL COMPANY EXECUTIVES I SAY WE NEED MORE MONEY
11/6 © TSS
WITH INCREASED PROFITS WE CAN EXPLORE—
—NEW WAYS OF MAKING MORE MONEY!
WICKS

There is no reason why that terrific burden of $1.685 billion in compensation payments on oil imported by the majors on the east coast couldn't be reduced by having a nationalized oil industry which could buy imported oil on a state basis until such time as Canada can develop the oil sands and offshore oil and bring in oil from the Arctic. Canada is going to have a difficult period of oil shortage from 1981-1991. During that period, Canada ought to import the necessary oil through a national petroleum corporation.

Napoleon said that an army marched on its stomach. Modern industrial nations march on fossil fuels. They need fossil fuels not just for smoke-producing industries but also for agriculture and fishing. The only reason that on the prairies a farmer can produce on 2000 acres where he used to be able to farm only 160-320 acres is mechanization, because of petroleum.

The future of Canada is going to depend on whether or not the energy resources of Canada are going to be developed for the benefit of the people who live here or whether the "gifts of nature" are going to belong to the American oil cartel.

In my opinion, the nationalization of our energy resources is vital to our future as an independent nation. I don't need to tell you that nearly a century ago when foreign power utilities threatened to become a gigantic monopoly in Ontario, Adam Beck and Premier Whitney of Ontario launched what is today Ontario Hydro. Adam Beck said that the U.S. takeover of electrical generation in Canada would result in a "political subservience that no Canadian can or will tolerate." I would remind you that the first publicly owned power corporation in Canada, Ontario Hydro, took as its motto, "The gifts of nature belong to the people." I submit that that's a better motto than Charlie Wilson's motto, "What is good for General Motors is good for America."

JOHN SMART

John Smart is an archivist in the Public Records Division of the Public Archives of Canada in Ottawa where many of the records of Canada's Crown Corporations are held. He is head of the Historical Researchers Collective Bargaining Group in the federal public service. Outside working hours John Smart coaches hockey and baseball and teaches Canadian history in the Carleton University summer school.

13

John Smart

The Public Enterprise Tradition in Canada

There is considerable sentiment in Canada in favour of public ownership of the oil and gas industry. For Canadians considering this alternative Canadian history has set some precedents and some lessons.

Public enterprise has in fact a long tradition in Canada and it is one of which Canadians should be proud. A recent writer on the public enterprise tradition in Canada concluded:

> The given is the sparsely settled Canadian Community next to the growing American giant. The only way it can keep within hailing distance economically is by large public investment, and Canada in theory not being a kingdom governed for the benefit of private clans, the natural form for this investment to take is public enterprise. Public enterprise is indigenous to the Canadian demography.[1]

Canadians, therefore, have often stood in this position before. Faced with the question of how to develop a resource or provide an essential service Canadians have most often depended on the private sector. But when questions of national policy or national survival have been at stake Canadians have opted for public enterprise. And that choice has been made hundreds of times. Canadians, in sum, far from showing themselves afraid of public enterprise have, during their history, embraced it readily.

The fifty-one per cent of those surveyed who favoured public ownership of the oil and gas industry in Canada in the Gallup Poll of November 1975 were in the mainstream of their country's history. A higher percentage of the economic life of Canada is now carried on under public enterprise than is the case for any other western industrial country. A higher percentage of the Canadian economy is publicly owned than in Sweden. Herschel Hardin estimates that fully one-third of the corporate assets of Canada outside of the financial institutions are in the hands of public corporations.

Following World War I and the development of new means of communication the Canadian Broadcasting Corporation, TCA (now Air Canada) and the National Film Board were created as public enterprises; all of these activities are privately controlled in the U.S. During World War II—again a time when national survival was at stake—no less than twenty-eight new Crown corporations were created. Between 1940 and 1955 a total of fifty federal corporations were set up. In Ontario about fifty Crown corporations and agencies are at work.[2] Those who look at the situation now with regard to oil and gas and suggest that we bring into play the public enterprise tradition have many precedents to look to in Canadian history.

Yet there are problems with this tradition and with its interpretation. Public enterprise has been more popular with the Canadian people than it has been with social scientists or with the Canadian business community. Centres of American influence and penetration in Canada have also been critical of the Canadian public enterprise tradition.

Those raised in the American mode of thought have two problems. First of all, public enterprise is not an American tradition. Americans, therefore, do not have the history and background to understand Canada's experience. Second, public enterprise has been so often undertaken in reaction to an American thrust that Americans tend to interpret such action as anti-American and threatening to them.[3]

Contemporary writing in Europe and the U.S. on the nature of the capitalist state regards it as indivisible; all its spending and activities are seen as contributing to the buttressing of the private sector. This view cannot explain the cases of Ontario Hydro and the CBC which I will discuss. Canada's degree of control by the United States is unique in the modern industrial world and as crises arise and as the Canadian-American relationship undergoes change precedents will be set. Students who look for the pattern elsewhere will miss seeing the one that does exist.[4]

Historically, Canadians—especially our leaders in business and politics—have too often avoided problems and responsibilities and taken the softer options. We live with the results of this historical trend and need to reverse it. Canadians must begin to interfere in their own affairs, to possess their own country and its resources.

Public ownership can be rewarding economically, can move a society towards a more democratic basis, and can be more creative intellectually. And it stands historically as the means which Canadians have chosen to preserve the society, culture and economy from disintegration at American hands.

Canadians in the past understood these simple facts better than we do now. The American penetration of the economy since 1945 has been accompanied by a penetration of Canadian culture which has fractured our historical understanding of Canada. Nearly twenty years ago H.G.J. Aitken, discussing the historical pattern of Canadian development, had no trouble seeing public ownership in context:

> Throughout Canadian development expansionism has been defensive in character. It has been part of a general strategy of containing the expansionism of the stronger and more aggressive economy of the United States and preserving a distinct political sovereignty over the territory north of the present international boundary. Each phase of expansion in Canada has been a tactical move designed to forestall, counteract, or restrain the northward extension of American economic and political influence. Primary responsibility for maintaining and strengthening this policy of defensive expansionism has fallen on the state.[5]

Writing in the same period the authors of a study on Crown corporations found it easy to refute the belief that public corporations are always the same as private corporations:

> The corporate form of organization invites the drawing of parallels between Crown corporations and ordinary limited companies. The people of Canada, it has often been stated, are the shareholders of Crown corporations, and members of the House are shareholders' proxies. Members have put this point of view forward particularly to justify their demands for information on the internal management of such large corporations as Canadian National Railways and the Canadian Broadcasting Corporation. A comparison between publicly owned and privately owned corporations shows, however, that the affairs of the former are submitted to far greater scrutiny than those of the latter. The annual general meeting of a privately owned company seldom lasts for more than an hour or so, and its chairman is part of the management. The affairs of a Crown corporation may be discussed in the House with no member of the management present, and in a committee of the House, with the chief executive officer appearing as a witness for days on end.[6]

The Canadian public enterprise tradition divides into two streams. One of these streams involves cases where the government has undertaken economic activity as a support to private business or has taken over essential economic activities which are no longer profitable. Public enterprise of this kind is not really public enterprise within my stricter definition; it is really just private enterprise activity leased back to the government. There is, however, another, richer stream of the public enterprise tradition in Canadian history. This involves cases where government has entered into economic activity as the result of popular pressure and in the interests of national survival. It is two cases of this kind I wish to discuss: the creation of Ontario Hydro and the Canadian Broadcasting Corporation.

In both cases public opposition to American activity within Canada grew to the point where the responsible governments were forced to act. It is possible to argue that in both cases the public corporations which were founded on the basis of public support have moved away from their original objectives and from their original constituency. Their subsequent history is less important, however, than their origins. In other words, it is more important to note that in previous situations public interest groups have achieved significant changes in government policy in response to American neglect of Canadian interests.

The quality of a public enterprise will depend a great deal on its roots in the society, on the way it is created. In the case of Ontario Hydro before World War I, a provincial public enterprise was created only after a long public fight which defeated two governments and led to the creation of a truly remarkable popular movement in the province.

The genesis of Ontario Hydro was as follows. It was clear for decades that hydro power for southern Ontario would have to come from Niagara Falls. By 1900, however, Ontario was behind Quebec in the degree of hydro development attained. American private power interests were developing Niagara under U.S. leases and had leases through Canadian front companies for developing power on the Canadian side. Yet the installations were being built only on the U.S. side and U.S. government and business interests were beginning to claim that seventy-five per cent of the power potential of Niagara should be developed on the American side on the grounds that their population was greater, their power needs greater, and that they had been first to develop.

Faced with the loss of the Canadian share of the greatest power site in the world, the people of Ontario became alarmed. An alliance of power consumers, manufacturers, municipal politicians, trade unionists and farmers came together after 1902 in what was called the public power movement. In 1905, a Conservative government pledged to public power was elected, bringing down a thirty year old Liberal regime which had granted the do-nothing leases.

Within a few months of the election, the Hydro Electric Power Commission of Ontario (HEPC) was created and the leader of the public power movement, Adam Beck, was installed as its Chairman. Beck was also taken into the Ontario Cabinet as minister without portfolio. These steps were seen by all as major victories but further battles remained. In reaction to the first onslaughts of the public power movement the declining Liberals had cancelled some of their American leases and had issued leases to a blue ribbon group of Toronto financiers and capitalists including leading men from the banks, railroads and breweries centred in Toronto. This new company—the Electrical Development Company—then took over the fight to keep power in Ontario in private hands.

Beck and the people behind him were more than equal to this challenge. The Toronto company and its allies proposed that the HEPC be restricted to a regulatory role comparable to the National Energy Board today; that it be an agency to license exports to the U.S. (which the Toronto outfit had contracted for) and to set prices while ownership and distribution would remain private. When that proposal proved politically unacceptable, the private company desperately lowered its proposed price for power by one half and offered to build all transmission lines under government inspection at cost.

A second "power" election in 1908 returned the public power Conservatives to office with a 66 seat majority and public power, chalked up another victory. The necessary legislation was passed and Ontario Hydro was in business.

On October 11, 1910 the first currents of public power over the Hydro lines reached Berlin (Kitchener) Ontario. Here is the account of the event provided by H.V. Nelles:

> The more conservative *Daily Telegraph* the next morning simply carried a full front-page portrait of Adam Beck and inside devoted itself fully to reporting the previous day's carnival happenings. Banners were strung between light standards already draped with bunting; thousands lined the streets to catch a glimse of Beck and [Premier] Whitney passing in open motor cars. Late in the afternoon the crowds jammed the dim Berlin arena for the real show. Hilda Rumpel, swathed in red, white and blue, carried the button, which would summon the electric genie from Niagara, atop a plush red cushion. Playing his part magnificently, Premier Whitney took the hand of Beck, his Minister of Power, and guided it toward the button. The crowd hushed as the Premier solemnly announced: "Gentlemen, with this hand, tried and true, this hand which has made this project complete, I now turn on the power." Suddenly the arena blazed with light, the current from Niagara surged through the mass with one great cheer. Little Hilda Rumpel's crown of tiny light bulbs glowed proudly, and the crowd roared itself hoarse.
>
> The mayor of Berlin thanked Beck effusively in German, and Beck responded by praising Berlin's own hydro pioneers, his own technical staff, the united

Adam Beck riding in parade, turning-on ceremonies, Berlin, Oct. 11, 1910

The turning-on ceremonies at Berlin, Oct. 11, 1910

Musn't Be Tricked Again

Public ownership of electrical energy means, first of all, cheap light and lots of light for all the people, including the poor people. Every little house can have lots of cheap light, and there is no such clean, reliable and free-from-work light as electric light. And as soon as the provincial distribution is built it means cheap light for every farmer in Ontario—a light in every bedroom, in his stables, in his yards, that lights without matches, that you can put out of commission the second you are thru with it.

Up to the present—except in those towns like Orillia, where there is public ownership—electric light has been dear light, and for the few.

The greatest light that God gave to man is the pure white light generated by God's greatest masterpiece—the Niagara Falls. Do not let the middlemen—the Gibsons, the Pellatts, the Jaffrays, the Nichollses—get between the people and this great blessing and make it dear and limit its use, so that they may be rich.

Let us keep it forever for all the people, and let us put it in the house of every citizen, however humble, at cost price. We want no electric barons here, as we now have the coal barrons of Pennsylvania, tolling us and tithing us at $7 a ton for coal, that public ownership of mines and of railways would give us for $2.50 a ton, or less.

We have been let into one trap; let us avoid the other, now spread for us.

The Toronto World, Nov. 21, 1907.

A Lesson from Socialists

It is possible that we have a Government in Ontario which does not understand that faith must be kept with private investors and that the public honor must not be tarnished. Is a Conservative Government, the hereditary guardian of the rights of property, to persist in a course of action which is admitted to be unjust even by British Socialists? The question comes with equal significance to the Council and ratepayers of Toronto. This city must continue to make heavy borrowings in London and it cannot afford to engage in confiscatory enterprises. So, many of the municipalities of Ontario must go from time to time to the London money market, directly or through private companies organized to establish public utilities, for necessary accommodation, and it is of vital consequence that their financial reputation should not be impaired.

The Toronto News, Dec. 12, 1907.

municipalities, and the Whitney government. "You say our work is done. I say that it has only begun. We must deliver power to such an extent that the poorest working man will have an electric light in his home. No more oil and gas, and soon, I hope, no more coal." . . . When his time came Whitney briefly recapitulated the tremendous campaign that had been mounted to crush the government's power policy: "We have been attacked, vilified and even slandered." Men of influence, he claimed, "from the humblest man in the land up to the Prime Minister of Great Britain were approached in the endeavour to destroy our power legislation. . . ." And at what was the ending of one struggle and merely the beginning of another, the Premier mustered a moving word of praise for Beck: "We, Adam Beck's colleagues, can never forget his steady confidence in the result and the bravery and pluck with which he stood up against all attacks." It was Beck's day, and the Premier stayed contentedly on the sidelines. It was also Beck's day when the morality play was staged again and again in the nine other participating hydro municipalities.[7]

When Beck said that the work had just begun he was quite correct. In 1910 Ontario Hydro was only a distribution system with the power to set prices. It was another 15 years before Ontario Hydro became a major producer of its own power. During the next years, Beck, who continued as Hydro Chairman until his death in 1925, worked to build popular acceptance for publicly owned generating stations.

The fuller story can be read in H.V. Nelles' book but Beck's career and personality illustrate some important points about the public enterprise tradition.

It is often said that private enterprise is the milieu in which truly interesting, truly daring, and creative people move. Let us look at Beck's career. Before public power he was a cigar box manufacturer in London, Ontario and well-off. He found that life very limited and became a popular Tory mayor of the city. He came to prefer public service. Once in provincial politics and a member of the Cabinet he never forgot why he was there nor who his constituency was. He refused to obey the rules of Cabinet and party solidarity when they conflicted with the objective of public power. Whenever there was any backsliding in the government on public power he took the issue to the people and won. He organized demonstrations, mass meetings, petitions and speaking tours and kept people informed fully on what private power interests were doing.

His last great battle—and this one he lost—was in opposition to the adoption of the American automobile in Ontario as a means of transportation. Beck's preference was to use public hydro power to fuel an inter-city system of radials—electric trains to carry the population to and from work and between cities. In other words, fifty-five years ago, a

Crown corporation in Ontario was the source of some truly creative thinking about the automobile and its long term utility as a means of transportation. When the Conservative Cabinet would not back the scheme he resigned and ran as an independent Tory in the 1919 election in which the government was defeated.

The farmer-labour coalition of the United Farmers of Ontario which formed the new government even considered taking Beck as their leader (the farmer leader having been defeated in the election).

Beck's failure was an important one in the history of Canadian politics and needs more examination. Beck and the UFO came close to an agreement. The labour representatives would have preferred him to the final choice, E.C. Drury, who proved unable to cope with the problems of the new government which, consequently, lasted only until 1923 and could not get re-elected. Beck's terms were that he wanted to set government policy on two questions: radials, not highways was to be the government's transportation policy; and Ontario Hydro was to be left as a commission under the direct control of its chairman. The UFO were skeptical of radials as essentially serving city dwellers, and not farmers. They also wanted to create a Ministry of Power in order to bring Hydro under the direct political control of the Cabinet and Legislature.

From the vantage point of the 1970's we can say that Beck was correct on the radials question but that the farmers were correct that principles of representative government demanded that Hydro policy be brought fully under the control of elected representatives. Beck's experience with the Conservative Cabinet between 1905 and 1919 had made him skeptical of the ability of politicians to resist private power's arguments and money. The public radio advocates of the 1930's had similar (and mistaken) attitudes and opted for commission status for the CBC. Attractive in the short run, this option in the long run removes a government activity from public control but does not remove it from government control.

There was a failure of imagination on both sides which has had a long term significance in the politics of English Canada. The post-war situation in Canada clearly demanded a coming together of two distinct creative elements—the nationalists of the Conservative party and the new class conscious elements of farm and labour represented in the UFO in 1919 and the Progressives in 1921. Only that combination had any positive hope of developing a new National Policy in the 1920's as an alternative to the policy of acceptance of American penetration on which the federal Liberals had stood since 1896. Had Beck become Premier in 1919, a new political party would have been formed in Ontario combining the Conservative nationalists of the public power fight, political labour and political farmers. Such a step in Ontario would also have had great effect on national politics. This missed opportunity nevertheless illustrates the remarkable innovative qualities Beck possessed.

The public enterprise which Beck founded went on to lead the world in power production and in hydro technology. It also attracted the constant criticism of the private power interests in the United States who were afraid that Ontario Hydro's success would be attractive to American state governments. U.S. private power regularly intervened in Ontario politics with money and propaganda. Beck responded to their efforts with a clarity and frankness which U.S.-Canada relations has unfortunately lost; that is to say, he fought back.

When we turn to consider the creation of the Canadian Broadcasting Corporation we see certain parts of the Ontario Hydro experience repeated. The first radio broadcasting licence in Canada was issued to the Marconi Company in 1919. The public radio movement did not develop in strength until nearly ten years later. After nearly a decade of allowing a free field to private enterprise radio in Canada, Canadians were sufficiently horrified by the results of the Americanization of Canadian radio to demand changes that could only be achieved through public enterprise.

By 1929 there were only sixty-two stations in Canada with only 300,000 licensed listeners in the major urban audiences. Only about sixty per cent of all Canadians had access to Canadian stations but virtually 100 per cent of Canadian listeners could hear American stations. NBC had affiliates (network stations) in Toronto and Montreal and was intending to open others. CBS had announced its intention to follow NBC into Canada in order to compete with its American rival for Canadian audiences.

Canada, a hinterland of American radio, was underdeveloped and was being cheated out of a major new form of communication. Large portions of Canada were being neglected in favour of concentration on southern Ontario; in particular, Quebec and the west were badly served by private radio. A 1925 survey showed that seventeen out of the twenty most popular programmes were American. Lack of government activity was allowing other major problems to develop. Canada had been short-changed on the allotment of station bands on the dial and the American government had not been effective in dealing with cases where American stations were located on Canadian bands and interfered with Canadian stations. Radio in the U.S. was already dominated by CBS and NBC; these two networks controlled seventy-five per cent of all available American radio frequencies by 1932. Large private companies along with CBS and NBC were branching out into radio in Canada and listeners feared the same unhealthy concentration would develop in Canada.

In December, 1928 the federal government took its first step towards public radio. A Royal Commission was appointed to inquire into radio in Canada. The chairman was Sir John Aird, Chairman of the Canadian Bank of Commerce, and the other commissioners were Charles Bowman, editor of *The Ottawa Citizen* and Dr. August Frigon, Director of Technical Education for the province of Quebec.

In its report the Aird Commission recommended that radio be taken under public ownership in Canada and stated that a continuation of private radio was likely to lead to American domination of the field in Canada. "The Aird Report was solidly based on the sentiment of Canadian nationalism and reflected a noble concept of broadcasting purpose and potential."[8]

In order to fight for legislation based on the Aird Report, public radio advocates came together to form the Canadian Radio League in 1930. Two Ottawans, Allan Plaunt and Graham Spry, supplied most of the executive leadership of the League and they brought together an impressive collection of activists and supporters who were able to out-lobby the private broadcasters and win the legislation (from the Conservatives in 1932 and the Liberals in 1936) which set up the CBC. When the fight for public radio did develop Imperial Tobacco, Robert Simpson Company, Swift Canadian, Quaker Oats, Borden's, Pepsodent, Philco, Dominion Stores, Wrigley's Gum, *The Toronto Telegram* and the Canadian Pacific Railway all contributed their names and money to try to keep radio in Canada in private hands.

Graham Spry, who went on to become, briefly, a leader of the CCF in Ontario, and later a Standard Oil executive, saw the radio issue very clearly. At one point in the fight he asked:

> Why are the American interests so interested in the Canadian situation? The reason is clear. In the first place, the American chains have regarded Canada as part of their field and consider Canada as in a state of radio tutelage, without talent, resources or capacity to establish a third chain on this continent.... In the second place, if such a Canadian non-commercial chain were constructed, it would seriously weaken the whole advertising basis of American broadcasting: The question before this Committee is whether Canada is to establish a chain that is owned and operated and controlled by Canadians, or whether it is to be owned and operated by commercial organizations, associated or controlled by American interests. The question is, the State or the United States?[9]

The American networks and the Canadian private stations wanted the choice to be made in favour of the United States, not the State. As part of its campaign against public radio NBC carried a one hour special programme devoted to Canada over its entire U.S. network in order to demonstrate its interest in the country. *The Toronto Telegram* cooperated with NBC on the project and the programme was much publicized in Canada. At a more effective level, Canadian private stations began to carry the most popular American shows as part of their own programme schedule. This was an attempt to undercut the Canadian Radio League's point that the status quo forced Canadians to listen to U.S. stations.

In spite of these activities on the part of American and Canadian private radio, other facts bore more weight with the federal government and Bennett and King were forced to choose the State, not the States. In 1931 the traditional *King's Broadcast* could not be carried in Canada because American Telephone and Telegraph, and Bell Telephone, who controlled the necessary cable systems, refused permission. R.B. Bennett was shocked to find that the degree of American control of Canadian radio could have such consequences. Legislation setting up the Canadian Radio Broadcasting Commission followed in 1932. In addition to the popular support organized by the Radio League, the governments of the three prairie provinces formally endorsed the Aird Commission report and asked for public radio legislation.

The problem with the legislation of 1932 and 1936 was that the federal government did not make the CBC a monopoly by buying out the private stations then in existence. Instead, a greater number of new private stations were licensed than the total created from the new CBC network. The CBC was supposed to regulate all broadcasting in Canada but it did not control it. In simple terms, the CBC missed part of the experience of Ontario Hydro where Beck went on to fight for and achieve monopoly for his public corporation. The CBC was stopped half way between the National Energy Board model of regulation only, and the Ontario Hydro model of monopoly control.

In discussing the 1936 Act, R.B. Bennett (then out of office) at least identified correctly the problem and the mistake made by his government:

> Owing to lack of funds, our organization was not able to build stations. It had to utilize by leases and other arrangements, the existing facilities for the purpose of carrying out its undertaking. There was a constant endeavour on the part of private interests to secure licences for new stations . . . All you have to do is grant enough licences and you destroy the public character of Canadian broadcasting. At the moment we have secured the ownership of this facility, or utility, or whatever you may call it, in the people of Canada, and we have ensured it against destruction by private interests, but the private interests have been

very vigilant. They have been watchful of their own interests and have determined that if it is within their power, this facility is not to function as a public enterprise. They have determined fairly and legitimately, so far as that is concerned, to destroy this publicly owned facility. Their idea is that it should be in the hands of private enterprise. The two or three great enterprises across the line have watched with increasing anxiety the operations of this publicly owned facility. They do not desire that their privately owned enterprises should be regarded merely as tentative experiments; they wish to continue as a privately owned enterprise and not be merged in a publicly owned and controlled broadcasting system such as we have in this country.[10]

The struggle for the CBC, therefore, only partially resolved the question of the State versus the States in Canadian broadcasting. The historical compromise with the private Canadian broadcasters left in place a pipeline for American programming in Canada. The traditional underfunding of public broadcasting in Canada did not allow the CBC to compete on equal terms for the Canadian audience. Yet, had the CBC not been created the continuing damage to the culture and identity of the Canadian people would have been incalculable.

The CBC's best programmes reached a mass audience in Canada with listening ratios higher than those of the American programmes. And the CBC repeated the experience of other Crown corporations in that its employees proved more creative and innovative than their conterparts in the private sector—as the CBC's record of international radio awards shows.[11]

In conclusion, the public enterprise tradition in Canada operates in the following historical setting. When U.S. penetration sets up a threat to national survival, the Canadian reaction is to turn to public ownership as the alternative. Canadians turn to public ownership only after a long period of neglect of Canadian interests. This was true in the cases of Ontario Hydro, the CBC, and also of the National Film Board and Trans-Canada Airlines (now Air Canada). In all these cases, American enterprise was given a free run in the Canadian field until the results could be supported no longer. Obviously, today a majority of Canadians believe that we have reached that point with oil and gas.

There have been problems with public ownership and publicly owned corporations but they have been more manageable than problems created by continued U.S. private ownership. The traditional criticisms of public ownership fail to cover the Canadian experience because they fail to see the root cause of public ownership in Canada—national expansion and survival. These criticisms are rooted in other societies—notably social democratic Europe and capitalist United States where national survival has not been a factor in the creation of public enterprise.

Public enterprises in Canada have performed well and have been centres of technological innovation. During the 1920's Ontario Hydro solved technical problems in hydro transmission which made it a much copied leader in the field. The National Research Council, CBC, NFB, Connaught Laboratories, even Air Canada and CMHC, have all had their successes and at various times have been leaders in their fields.

It is also clear that the public enterprise tradition has been weakened since 1950. CP Air, CTV and cable TV have been allowed to enter fields where public enterprise previously had a monopoly. The Canadian Radio Television Commission has taken over from the CBC the responsibility of regulating all broadcasting in Canada. The CBC now undergoes periodic licence renewal examination by the CRTC just as CTV does. Recently created Crown corporations like Petro-Canada and the Canada Development Corporation are expected to compete in the American-dominated Canadian economy on the same basis as other corporations but without equal resources.

It is clear to me that nationalization of Canadian oil and gas would certainly be within the Canadian public ownership tradition and that that tradition is strong enough to give Canadians confidence that such a step could be taken successfully. It is also clear from that tradition that a healthy controversy and a real fight over the question could be expected. Those interested in the continued private ownership of Canadian oil and gas would certainly put their case strongly before the Canadian people.

The development of the public enterprise tradition has been closely related to efforts for national survival. It is fitting to conclude with reference to two historical episodes that demonstrate the continuing urgency of the issue of national survival.

Mackenzie King ended his long career as Prime Minister in 1948. When he was turning over the government to the younger men, St. Laurent and Pearson, he felt it necessary to express very strongly his forebodings about the United States in Canada's future.

He had told the Cabinet in 1946 that he "believed the long range policy of the Americans was to absorb Canada" and that "it might be inevitable for us to have to submit to it—being so few in numbers and no longer able to look to British power for protection."[12] By the time of the actual transfer of power in 1948 King had decided that absorption was not necessary and must be resisted. He wrote in his diary:

> I told Pearson that while I might miss to be the head of the government, I would never cease to be a Liberal or a British citizen and if I thought there was a

danger of Canada being placed at the mercy of powerful financial interests in the United States and if that was done by my own party, I would get out and oppose them openly.[13]

Mr. Pearson received that advice and went on to have his own career as Minister of External Affairs, Leader of the Opposition and Prime Minister. When he came to retire from the latter office in 1968 he felt it necessary to voice similar warnings to his successor:

The industrial and economic and financial penetration from the south worries me, but less than the penetration of American ideas, of the flow of information about things American; American thought and entertainment; the American approach to everything.[14]

There has clearly been too long a tradition of Canadian Prime Ministers leaving office expressing concern about the American danger to Canadian interests. A judicious return to the public ownership tradition might well save a third modern Canadian Prime Minister from having to express similar regrets.

1 Herschel Hardin, *A Nation Unaware,* Vancouver: J.J. Douglas, 1972, 76.

2 Peter Silcox, "The Proliferation of Boards and Commissions," in T. Lloyd and J. McLeod, *Agenda 1970,* Toronto: University of Toronto Press, 1968.

3 U.S. Ambassador to Canada Thomas Enders is well within the American tradition when he protests on behalf of his government the public ownership of Saskatchewan potash. Ambassador Enders recently complained to an American audience: "This is a classical example of how things could go wrong. . . . If Saskatchewan succeeds in taking over the industry, it could be a precendent for other provinces to take similar action." Interview reported in *U.S. News and World Report,* June 21, 1976.

4 *See* Rick Deaton, "The Fiscal Crisis of the State in Canada," in D. Roussopoulos (ed.), *The Political Economy of the State,* Montreal: Black Rose Books, 1973; Ian Gough, "State Expenditure in Advanced Capitalism," *New Left Review,* July, 1975; James O'Connor, *The Fiscal Crisis of the State,* New York: St. Martin's Press, 1973.

5 H.G.J. Aitken, "Defensive Expansionism: The State and Economic Growth in Canada," in W.T. Easterbrook and M.H. Watkins (eds.), *Approaches to Canadian Economic History,* Toronto: McClelland and Stewart, 1967.

6 C.A. Ashley and R. Smails, *Canadian Corporations,* Toronto: Macmillan, 1965, 19

7 H.V. Nelles, *The Politics of Development,* Toronto: Macmillan, 1972, 301-2.

8 Frank Peers, *The Politics of Canadian Broadcasting 1920-51,* Toronto: University of Toronto Press, 1969, 51.

9 *Ibid.*, 91.

10 Canada, House of Commons, *Debates,* June 15, 1936, 3710.

11 At least one young producer at CBC radio knows some of this history and has been inspired by it. Krista Maeots, now responsible for CBC Radio's national morning show said recently:

> Listening to programmes in the CBC archives, I'm infatuated with the sound of the thirties and forties, especially the 1936 Christmas show, which moved from one place to another across the country, live. There was a tremendous sense of excitement. I also love the live drama of that period. . . .
>
> The major impetus of my programme will be to bring the country into the show, live. We'll have interesting people interviewed live on the air from studios all over the country, using the best facilities for good radio quality. I think that's the only way the programme can be really national.

Reported in *The Globe and Mail,* July 21, 1976.

12 Quoted in C.P. Stacey, "The Turning Point: Canadian-American Relations during the Roosevelt-King Era," *Canada: An Historical Magazine,* Autumn, 1973,8.

13 *Ibid.*

14 Quoted in Frank Peers, "Oh Say, Can You See," in Ian Lumsden (ed.), *Close the 49th Parralledl etc.,* Toronto: University of Toronto Press, 1970, 136. Pearson was speaking to a Canadian Press reporter.

WALTER L. GORDON

Walter L. Gordon is Chairman of the Board of Canadian Corporate Management Company Ltd. Elected to the House of Commons as Liberal MP for Toronto-Davenport in 1963, he served in the government as Minister of Finance and President of the Privy Council. He is a Director, Toronto Star Limited; Chancellor, York University; Chairman, The Canada Studies Foundation (1970-75); and was one of the founders of the Committee for an Independent Canada. His latest book is *Storm Signals—New Economic Policies for Canada* (1975).

14

Walter L. Gordon

Regaining Control of Canada's Resources

To begin with, I should state my personal position on the debate over Canada's national resources. It is not an ideological one. I am not a socialist; not a member of the NDP. However, I believe it is bad for Canada to have so much of its natural resources and so many of its business enterprises controlled by foreigners. I am a member of the Liberal Party, but I do not believe the members of the present Liberal Administration in Ottawa have tackled seriously the question of foreign control of the economy. They are a long way behind public opinion in this matter. As I am not a socialist, I suppose it can be said that I am a supporter of the capitalist system. That is true up to a point: I believe firmly in the advantages of decentralization in most phases of day-to-day decision making.

However, despite the present mood in the country of resistance to change, despite the feeling in all political parties that they should be moving to the right and despite the many mistakes that have been made and are being made in Ottawa about economic policy, I believe we must look forward to more government intervention in the future, not less. I say this because with an increasing concentration of power in the hands of Big Business (much of it controlled abroad) and in the hands of the Big Trade Unions (many of them influenced from abroad), there must be a countervailing force. That force can only be Big Government which at least, in the last analysis, is controlled at home.

If, therefore, I am to be labelled a capitalist and a free enterpriser, it is only with decided reservations. I expect there are some self-proclaimed free enterprisers—incuding some who still live in the nineteenth century—who may not agree with me at all. For example, I do not trust the motives or the policies of the international oil cartel or their representatives in this country. At the same time, I do not believe the Ottawa bureaucracy is equipped to manage the Canadian oil industry, or any other industry, in any detailed way. On balance, my preference would be to have a federal Crown Corporation direct the affairs of the oil industry in Canada rather than leave it in the hands of international corporations whose practices and objectives are open to considerable question. I do not feel the same way about some of the other resource industries.

It may be useful to state very briefly why I believe it is against Canada's best interests to permit the present extensive foreign control of the economy to be continued. To begin with, it is a fact of life that in this country, as in the U.S., business leaders wield considerable influence over public policy. To the extent that these business leaders are the chief executives of Canadian subsidiaries of U.S. or other foreign corporations (as is the case with the large oil companies, for example), they may reflect the views and interests of such foreign corporations. It would be surprising if this were not the case. And these views and interests may not necessarily coincide with the best interests of Canada. In the case of the oil industry, they have not always done so in the past.

I believe that our "branch plant mentality", occasioned by so much foreign control of Canada's industry, has a stultifying effect on Canadian managers, scientists and technicians. This works to our disadvantage. Canadian subsidiaries frequently are required to rely on their foreign parents for research and new product development, which is restricting to Canadian managers, scientists and technicians. Key decisions such as the expansion of markets, the location of new plants, and the promotion of senior personnel are made at head office, usually in the United States. This is not the best way to develop management and scientific talent in Canada.

In the manufacturing field, foreign-controlled companies tend to import parts and components from their parent corporations and associates instead of developing alternative sources of supply in Canada, thereby creating more employment opportunities here. Despite the opinion of the Chairman of the Economic Council of Canada, most people, certainly most provincial governments, would like to have more, not fewer, secondary manufacturing establishments in their regions. Moreover, in the manufacturing sector, Canadian subsidiaries of foreign parent corporations frequently are restricted from developing export markets for their products. Again, this has an inhibiting effect on employment opportunities.

The disadvantages of so much foreign control of the Canadian economy that I have outlined very briefly are now becoming understood by the general public in all parts of Canada. According to opinion polls and surveys, the public believes that something should be done about this. As previously noted, governments have not kept pace with public opinion on this issue. It is also well to remember that in Canada, as in the United States and other industrialized nations, industry has developed along monopolistic or oligopolistic lines. The result is that one or more companies in a particular industry tend to set the pattern for the industry as a whole, especially when it comes to prices. The earlier ideas of free competition do not apply any longer in these cases. In many of the industries in question, the leading companies are controlled abroad, giving them an even greater influence over the Canadian economy than would be the case under conditions of free competition in the old-fashioned sense.

One solution to the problem of foreign control of the Canadian economy would be a free vote in the House of Commons supporting the view that in the opinion of the Members, the Canadian public believes that the owners of the largest Canadian subsidiaries of foreign parent corporations should sell control to Canadians (not to the Canadian government) over a period of, say, ten years. I have suggested this should apply to the thirty-two Canadian subsidiaries with assets in excess of $250 million. This would include all the principal oil companies. The advantages of this proposal would be:

1. Only thirty-two companies would be affected.
2. The companies would not be nationalized.
3. No legislation would be required and there would be no need for sanctions.
4. It would be left to the foreign owners to decide how to go about selling the shares of their subsidiaries to Canadians.
5. They would have plenty of time to work things out.
6. For all practical purposes, the transfer of ownership of these thirty-two large companies would resolve the foreign control issue. The remaining seven or eight thousand smaller foreign-controlled companies in Canada would not be involved in any way.

The amount of capital to be invested would amount to about $1.5 billion a year, or less than one per cent of GNP—something well within Canada's financial capabilities.

In the light of this, I would like to touch very briefly on: the oil and gas industry; the projected gas pipeline from the Arctic; hydro electric developments; and other energy sources, including uranium and coal.

A few years ago, Canadians (including Canadian officials) were misled into believing we had unlimited quantities of crude oil to meet all possible requirements for the foreseeable future. In a speech to the Canadian Institute of Mining and Metallurgy on June 2, 1971, the then Minister of

Jerry McAfee: Profile of a Remarkable "Canadian" Corporate Citizen

The road to the top of Gulf International has taken Texan Jerry McAfee through a many faceted career that included a six year stint as president of Gulf Canada Ltd.

Beginning his career with Gulf in Texas in 1945, he climbed the ladder of the corporation with a posting in Britain in 1964 and was appointed head man for Gulf in Canada in 1969.

In Canada, McAfee's status as the foreign head of a branch corporation did not prevent him from making frequent pronouncements on Canadian energy policy. When US Senate hearings revealed that American oil companies had made large political contributions to the Liberal and Conservative parties in Canada during the 1970s, including $140,000 a year by Gulf, McAfee defended the contributions. He made it clear that his company would continue to make such contributions in Canada to any political party that did not include the nationalization of petroleum companies in its platform.

Throughout his stint in Canada, McAfee kept his American citizenship. It came in handy early in 1976 when a tumultuous shake up at Gulf International resulted in the resignations of four senior executives, including chairman and chief executive officer Bob R. Dorsey. Gulf's management had been torn apart by revelations of the company's questionable political contributions both in the US and abroad.

With the old team in ruins Gulf turned to their man in Canada, Jerry McAfee. Bidding Canada farewell, McAfee took the helm at Gulf International.

McAfee is missed by the Canadians who knew him. On hearing of his departure, W.O. Twaits, former chief executive officer of Imperial Oil, remarked: "We need all the wise brains we can get in this country. We're sure as hell going to miss him."

McAfee is also missed by the many associations he joined while in Canada. He belonged to: the National Council of the Canadian Manufacturers' Association; the Toronto Club; the Canadian Club of Toronto; the Board of Trade of Metropolitan Toronto; the Empire Club; the Queen's Club; the Rosedale Golf Club; the York Club; and the Royal Canadian Yacht Club.

MISTER
CLEAN

Energy, Mines and Resources stated: "Canada's total oil reserves were 469 billion barrels at the end of 1970 while total natural gas reserves were 725 trillion cubic feet. At 1970 rates of production, these reserves represent 923 years' supply for oil and 392 years' for gas."

In a speech at Pittsburg on January 22, 1976, the present Minister of Energy, Mines and Resources, Mr. Gillespie, stated that Canada's gas and oil supplies were declining and that increasingly we would have to look to the international markets for oil. He continued: "As a result, we will be faced with a huge balance of payments deficit. Assuming no increases in OPEC prices by 1985 or sooner, this oil deficit alone could reach a staggering five billion dollars—every year." In the same speech, the Minister stated: "Natural gas prospects are almost as bleak. Spot shortages are predicted starting late in this decade. By early in the 1980's, a gap will appear between domestic demand and conventional gas supplies."

Both these ministers presumably obtained their estimates—in the early seventies very optimistic, now highly pessimistic— from the National Energy Board. Presumably, the latter body obtained its figures from the oil industry, that is from the Canadian subsidiaries of the international cartel. If either set of estimates were thought to be wrong, one would have expected the oil industry to point out the error. This was never done to my knowledge. The only logical conclusion to be drawn is that the estimates supplied by the industry were either very wrong in 1970 and 1971 or they are very wrong today. In these circumstances, how much credence should be placed on the estimates or pronouncements of the oil industry? I suggest not very much.

Incidentally, some well-informed individuals do not accept the current pessimistic views. On the contrary, they tend to think that in the foreseeable future there may be surpluses in the world of crude oil production. Who and what are we to believe?

I suggest that any of you who may think the operations of the international oil companies and their subsidiaries in Canada are motivated by altruism read: *The Seven Sisters* by Anthony Sampson; *The National Interest* by Edgar Dosman; *The Tar Sands* by Larry Pratt; *Canada's Energy Crisis* by James Laxer; *Sell Out* by Philip Sykes; and some perceptive articles on the subject by David Crane of *The Toronto Star*.

The impressions these books and articles left with me are that the descendants of the old pirate class in the early days of oil development are not so very different from their predecessors; that huge profits are the sole and overriding objective of big business, certainly the oil business, and justify everything (including bribery on a large scale in the case of Gulf); and that the gulling of government officials and the public can become a high form of art.

When the oil industry in Canada was pressing for increased exports, it was claimed the estimates of future Canadian supplies were astronomical. When, some three or four years later, the industry was pressing for much higher prices and thus higher profits, allegedly to finance greater development, it was claimed the estimates of future supplies were dangerously low. If the oil men who made these estimates in the first place and convinced the government officials of their accuracy were wrong, then surely they are not the men we should place our faith in.

Perhaps it was only coincidence that the very high and then dangerously low estimates supported the objectives of the industry at each point in time. It would be interesting to know. As to the alleged independent position of Canadian subsidiaries of the international companies, let us all consider what Larry Pratt has to say about this in *The Tar Sands*:

How much automony of decision making power does an affiliate such as Imperial Oil enjoy? The question is important since Imperial's executives have always taken pains to argue that Exxon/Jersey exercises little or no direct control over their major investment decisions, pricing policies, marketing strategies, and so forth. . . . Given the tightly knit nature of the international oil industry, this has always seemed an unlikely proposition; but until the Nova Scotia Power Corporation launched a $100 million lawsuit against Imperial over a fuel oil supply agreement, the evidence was unavailable to disprove it once and for all. The court case between the NSPC and Imperial, heard before a justice of the Nova Scotia Supreme Court in the spring of 1975, ended any remaining doubts. While the case centred primarily on the issue of whether Imperial was entitled to pass on to its customer, the power board, the burden of tax increases imposed in Venezuela, the evidence revealed much about the way Exxon directly involves itself in Imperial's pricing and crude purchasing policies. 'The judges found,' *Oilweek* reported, 'that Imperial does not have independence of action in deciding where its crude supplies come from or an ability to bargain directly with the supplier. This function is carried out by the planning and analysis group of Exxon International in New York.' Imperial has no representation among this small group, thus the company cannot control its own source of crude supply. 'The ultimate decisions rested with Exxon Corporation.'

Imperial has always been the pacesetter in the pricing of Canadian crude and refined products, but the Nova Scotia lawsuit demonstrated that this function too is determined by Exxon. Among the 40,000 documents assembled for the case was a Telex message sent from Exxon International in June of 1973 instructing the company's worldwide affiliates to increase prices and advising them of the desirability of making further 'market adjustments.' Before Ottawa stepped

Imperial Oil: As Canadian as ...

INTERVIEWER: But Mr. Armstrong, if it came down to the crunch with Exxon, if it got down to out-and-out disagreement, to bare knuckles, what then?

J.A. ARMSTRONG: If it got down to bare knuckles? Exxon would either go along with Imperial's executive committee and with me as Chief Executive Officer, or they could get rid of us. It's that simple.

...

J.A. ARMSTRONG: ... But above all, we're a Canadian company. Our capital is raised in Canada, and the decisions on how it is used are made in Canada. We are as Canadian as if our major shareholder were in Manitoba.

in to freeze prices in September 1973, crude prices had increased by 95¢ a barrel, or over thirty percent in less than a year; and each price rise was initiated by Imperial, no doubt as part of Exxon's global policy.

This evidence effectively lays to rest the lingering myth of Imperial's autonomy; neither it, nor any other foreign-controlled company operating in Canada, enjoys independence of action, particularly in areas such as the tar sands where billions of dollars are at stake. The companies behind Syncrude are American—none other. Syncrude is about as Canadian as the Stars and Stripes.

Since Mr. Pratt wrote the above, officials of Imperial Oil have acknowledged that Exxon/Jersey must approve all expenditures over $5 million. This demolishes any illusions that Imperial is independent and autonomous.

After reading the books referred to and thinking about what has happened, I, as a not overly sceptical individual, am not prepared to accept pronouncements by spokesmen for the international oil companies at face value. If they have been so very wrong in the past, can they not be wrong again? In these circumstances, I believe control of the industry in Canada should be transferred to Canadian hands. The best way of doing this in my opinion would be along the lines of the formula I suggested earlier: selling control of thirty-two large subsidiaries to Canadians. If that solution is unacceptable, then I would like to see the Canadian operations of the big foreign oil companies (or at least Imperial Oil, by far the largest of them) taken over by the government and operated in the best interest of Canadians. The smaller Canadian-controlled companies should be left alone.

The cost of acquiring the shares of Imperial Oil would presumably be related to their quoted market value. There are 130,188,546 common shares outstanding. At the closing market price on July 20, 1976 of $23.50 per share, this produces a total valuation of $3,060,000,000. Exxon Corporation of New York holds approximately seventy per cent of the outstanding shares of Imperial Oil which, on the same basis of valuation, would amount to $2,142,000,000. It follows that if the government purchased only the controlling block of shares from Exxon Corporation, the cost would be something over $2,000,000,000. If all the shareholders were given the right to sell on the same basis (as I think they should be), the total cost would be slightly over $3,000,000,000.

If nationalization proves to be the answer (and I like my first proposal better), I do not mean to suggest that this should become a new field of endeavour for the Ottawa bureaucracy. On the contrary, I would suggest the operations of these companies be directed by a Crown Corporation, possibly owned jointly by the federal government and the provincial governments most directly concerned. I suggest that this Crown Corporation might be managed by a Chairman whose background is finance, and a Canadian president who is a thoroughly experienced oil man. Alternatively, the government's new enterprise, Petro-Can, might be the answer. But in an undertaking of this size and importance, I would like to see it headed by two men of different backgrounds, not by a single individual.

If neither of the proposed solutions is acceptable for political or other reasons, then I would urge the establishment of an independent Royal Commission to enquire into all phases of the energy problems that confront this country. Such a Commission should be required to report and comment on what has happened in the past including pricing policies, tax and royalty policies, the ways in which estimates of reserves were arrived at and then changed, the influence that the propaganda of the oil companies has had on government policy, the profits earned and to be earned by the oil companies, etc. The terms of reference should be as broad as possible. In addition, such a Royal Commission should be required to make recommendations as to how Canadians can be assured of adequate supplies of energy in the future at the lowest possible cost.

If the government is not willing to require that the ownership of the Canadian subsidiaries of the international oil companies be transferred to Canadians, or not prepared to take over the companies in question, then nothing short of a full-dress enquiry by a Royal Commission should be accepted by the public. The subject is much too important for the future of our country for Canadians to be content with less. It is time the true facts were laid before us by some impartial body composed of individuals who can be trusted. In the meantime, no further increases in the price of oil or gas should be permitted.

What would be the effect of an increase in the price of oil? If the increase were to go to the oil companies in an attempt to persuade them to do more exploration work, I suggest the decision is naive. The international oil companies and their subsidiaries in Canada should be dealt with from a position of strength and not weakness. If the government is unhappy about the present levels of new exploration and development, it should take over the companies in question and get on with the job itself through the medium of a Crown Corporation.

If part of the increase in price is to go to the federal or Alberta governments in the form of increased tax revenues or royalties, then the public should be excused if it feels affronted. Canadians are being asked to accept certain disadvantages (I do not like the word sacrifices) under the Anti-Inflation programme. I believe we should all support the programme even though it may affect adversely our personal interests and even if some of the regulations need to be changed—especially those that will mitigate against efficiency and a reduction in costs. But the public can hardly be expected to continue to support the programme if the government allows any further increases in the prices for oil and gas. Such increases would obviously increase the cost of living. A suspicion that they are intended to placate either the international oil industry or the government of Alberta is hardly reassuring.

The next related matter to consider is the projected gas pipeline up the Mackenzie River Valley, and other possible gas and oil pipelines from the Arctic. The project being sponsored by Canadian Arctic Study Limited seems to be favoured by the federal authorities. In fact, there are those who allege that the authorities have aready decided in favour of this project despite the fact that the necessary hearings before the National Energy Board have barely begun. The thing that troubles me as much as anything is that the controlling interests behind this project are essentially the same international oil companies or their subsidiaries in Canada we have been discussing. Once again, let me repeat, I question the motives of these companies.

Turning to hydro-electric development, I think it can be argued quite logically that it makes sense to harness the resources of our northern rivers, provided the rights and livelihood of the native peoples are protected. I am more sceptical, however, about the argument that the electric power so generated should be exported except in off-peak periods. It seems to me this ties the Canadian economy even more tightly into a continental framework. Moreover, once Canada begins to export hydro electric power to the United States on a large scale, the practicability of ever curtailing such exports to meet future needs in Canada becomes questionable. Those who favour such developments may claim that without an agreement to export the power in question, it would not be possible to finance these projects. If this is true, it might be best to defer the proposed developments until the power is needed in this country.

The federal government moved some time ago to control the export of uranium and to prevent the projected sale of one uranium producer to non-residents of Canada. These moves are to be applauded. If the National Energy Board has not already full powers to deal with all phases of the production and sale of uranium, it should have. Similarly, I believe the National Energy Board should have full power to deal with the production and exports of Canadian coal mines and the ownership thereof.

Our long-term goal should be to preclude foreigners from owning any energy sources in this country. I have deliberately concentrated my remarks on the oil industry. I make no apologies for doing so. In the first place, oil and gas are of much the greatest importance. And in the last analysis, if we fail to come to grips with these particular resources, it will not matter very much what we do about the others.

Public Petroleum Association of Canada

Statement of Purpose

The Public Petroleum Association of Canada is a non-partisan organization whose purpose is to advocate a national petroleum policy that serves the interests of Canadians.

The sponsors of the PPAC believe that at present no coherent national petroleum policy exists. The petroleum industry in Canada is controlled by subsidiaries of the major multinational oil companies. These companies operate throughout the world, serving foreign investors and foreign governments, responding to profit opportunities on a global basis.

In Canada the traditional policy of the oil companies was to press for the export of more and more Canadian oil and natural gas to the United States. Since 1973 the oil companies have drastically revised their estimates of Canadian oil and natural gas potential production (producibility) downward. They now warn that the country faces potential shortages of petroleum within a few years—unless the oil companies are guaranteed a much higher return on their investment as an incentive to exploration and development of sufficient new reserves to meet Canadian demand.

The petroleum companies are taking advantage of high oil prices internationally to press for higher returns on their investment in Canada. National industrial strategy which directly affects the job security of many Canadians is heavily influenced by the new producibility estimates of the oil companies since petroleum is a basic building block on which the entire economy is dependent.

We believe there is strong evidence to suggest that government regulatory bodies at the federal and provincial level have tended to follow the lead of the oil companies in setting petroleum policies and in making producibility estimates. In fact, it appears that the primary source of information for the National Energy Board concerning all aspects of the petroleum industry has been these multinational companies. This means that the petroleum companies have far too much weight in determining the economic climate within which Canadian industry and agriculture operate.

The PPAC believes that Canada needs:

1 **A public commission of inquiry, to be established by the federal government, to investigate the enormous change in petroleum producibility estimates made by the oil companies in recent years.** Such a commission, with broad powers to subpoena the records of the petroleum companies, should investigate among other things the relationship between the petroleum companies and government regulatory bodies like the National Energy Board. The commission should include critics of the oil industry among its members.

2 **An end to ownership and control of Canada's petroleum industry by the multinational oil companies.** A first step toward achieving Canadian control of the industry would be the establishment of government marketing boards through which crude oil and natural gas would be sold. As the next step the presently foreign-owned major oil and natural gas firms must be operated as public utilities, owned and controlled by the federal government and the governments of the petroleum producing provinces. The principle of publicly-owned energy industries has long been recognized in Canada, in the area of electric power utilities. The same approach applies to petroleum. Where a natural monopoly exists in the supplying of energy to industry and consumers, the community at large should receive maximum benefit by organizing the energy industry as a public utility.

3 **A policy of petroleum self-sufficiency as a key element in a balanced programme of industrial development.** The rapid export of Canadian oil and natural gas at low prices has been disastrous for Canada. It has forced us to consider the development of costly replacements for depleted reserves sooner than would otherwise have been necessary. Canada should permanently phase out the export of oil and natural gas as soon as possible. Canadian petroleum should not be priced in terms of a world price that is convenient to an international oil cartel. It should be priced in accordance with the actual costs of producing energy for Canadians and the actual costs of developing future energy supplies for the nation. Canadian petroleum policy should serve as an element in a national industrial strategy geared to promoting greater indigenous manu-

facturing based on the nation's resources. Such new manufacturing development should be centred as far as possible in the regions of Canada that now concentrate on extractive industries alone—the west, northern Ontario and Quebec, and the Atlantic provinces.

4 **A serious commitment to safeguarding the environment and ending energy waste.** Massive petroleum projects in the Alberta oil sands, in the Arctic and off the Atlantic coast bring with them grave environmental risks. Such projects must not proceed on the assumption that environmental catastrophes caused today can be dealt with by tomorrow's society. Energy conservation must become a central concern of Canadian economic planning. Our concern for developing Canadian manufacturing is not an advocacy of activity for the sake of activity, that takes no account of resource depletion. The problem of resource use is a Canadian problem, but it is also a global problem. Canada should commit itself to advocating resource conservation on a global basis.

5 **To recognize the rights of the native people of Canada.** Control of their land by northern natives is essential to the integrity of their society. Energy projects in the North must *never* proceed until the land claims of northern native people have been settled in a manner that is satisfactory to them. In our view, no compelling case for immediate construction of gas and oil pipelines in the North West Territories has yet been made. Most arguments for northern pipelines are based on the assumption of a continuing high level of natural gas exports to the U.S.

The PPAC intends to further its aims in the following ways:
1 By carrying out a programme of public interest research on developments in the petroleum industry in Canada.
2 By issuing public statements on petroleum policy.
3 By appearing at public hearings related to the petroleum industry.
4 By organizing public events in support of a new petroleum policy for Canada.

To co-ordinate its activities, the PPAC is raising funds to support a national office.

Suite 780 86 Bloor West Toronto, Ontario
(416) 961-2032

Public Petroleum Association of Canada

Sponsors:

Thomas M. Anthony
David Archer
John H. Archer
Margaret Atwood
Nick Auf der Maur
Gregory Baum
Jane Bigelow
Richard Bocking
Silver Donald Cameron
Donald Chant
Stephen Clarkson
John J. Clement
Donald Creighton
Greg Curnoe
James Eayrs
Ed Finn
Richard Gathercole
Dave Godfrey
M. D'arcy Goldrick
J. L. Granatstein
George Grant
Herschel Hardin
John Harney
Abby Hoffman
Mel Hurtig
John Hutcheson
Avrom Isaacs
Karl Jaffary
Pauline Jewett
Bruce Kidd
Margaret Laurence
Jim Laxer
Dennis Lee
James Lorimer
Dennis McDermott
Ian McDougall
Clarke Mackey
Kenneth McNaught
Peter Martin
Robin Mathews
W. L. Morton
Farley Mowat
Michael Oliver
Charles Pachter
Robert Page
Douglas Pimlott
Walter Pitman
Fred Pomeroy
Larry Pratt
Al Purdy
John Richards
Heather Robertson
Abraham Rotstein
Dennis Smith
Michael Snow
Charles Taylor
Bill Temple
Dave Treleaven
Peter Usher
Mel Watkins
James Woodford

Notes to the Illustrations

ii *Miller Services*
xii *Miller Services*
xiii This Imperial Oil Ltd. commercial message appeared on *Hockey Night in Canada* telecasts, CBC TV, throughout the autumn-winter, 1975-76.

2 *Public Archives Canada,* C-20350.

4 *Fortune,* May, 1976.

22 *Maclean's,* April 5, 1976, 11.

32 Reprinted with permission, *The Toronto Star,* May 24, 1976.

40 *Miller Services*

46 *Miller Services*

50 *The Toronto Star,* August 11, 1970.

58 Material compiled by Mel Hurtig.

80 *The Edmonton Journal,* Dec. 4, 1973; D.D. Lougheed, "Address to the NWT Chamber of Commerce," Yellowknife, June 9, 1976.

93 *Miller Services*

98 Ontario, Ministry of Treasury, Economics and Intergovernmental Affairs, *Ontario Budget 1975; Supplementary Actions to the 1975 Ontario Budget; Ontario*

Budget 1976; Economic Impact on Ontario of 1974 Oil and Gas Price Increases, Doc. No. FP-4178, April 9-10, 1975; *Oil and Gas Pricing and Trade Competition: An Ontario View,* 1976.

104 *The Calgary Herald,* Jan. 15, 1976.

112 S.R. Cook & E.H. Davenport, *The Oil Trusts and Anglo-American Relations,* 1923.

116 Reprinted with permission, *Toronto Star Syndicate.*

122 *Toronto Symphony News,* April/May, 1976.

146 Imperial Oil Ltd., *Annual Report 1972,* Annual Letter from W.O. Twaits; J.G. Livingstone, "Address to 1976 AIPE International Plant Engineering Conference," Montreal, June 2, 1976.

157 *Public Archives of Ontario*

166 Victor Ross, *Petroleum in Canada,* Toronto: Southam Press Ltd., 1917, 28-30.

176 *The Ottawa Citizen,* April 23, 1976.

179 *Public Archives of Ontario*

180 *Public Archives of Ontario*

212 The Canadian Institute of Public Opinion, *Gallup Poll,* No. 382, Nov., 1975.

216 Reprinted with permission, *Toronto Star Syndicate.*

224 *Ontario Hydro.*

225 *Public Archives of Ontario.*

242 *The Toronto Star,* Jan. 17, 1976; *National Reference Book,* 1971-72, 529-30; Picture, *Gulf Commentator,* Feb., 1976.

246 "Imperial and Exxon," *Imperial Oil Review,* 1976, No. 1.

250 *The Globe and Mail,* Toronto, April 29, 1976.